# THE DIALOGUES

# The Dialogues

*A Religionist*
*Talks To A Skeptic*

Serge Bourgaine

ORBIS
UNUM
PRESS

Library of Congress Control Number: 2020908323

ISBN 978-0-9994906-1-7 (ebook)
ISBN 978-0-9994906-4-8 (print book)

βopγηv02ρα

*À ma femme, pour les années passées*
*et celles qui nous attendent, avec tout mon amour.*

# Contents

# Prologue

*The scientific method is deliberately constructed to overcome a deep-seated human wish to assume that what we want to be true - what we claim to 'know' - really is true.*[1]
Ian Stewart (1945-)
British Mathematician

*A summer morning, 2018*

This day, as most days, Simplicio awakens to the peal of a bell, its metallic timbre resounding from the distant tower of a Franciscan monastery. The bell announces Lauds, an office of the *Liturgy of the Hours*, summoning the monks to sanctify their day with prayer. Simplicio derives comfort from the sound. He envisions the predawn ritual of the brothers, each reaffirming the constancy of his faith, each reasserting the truth of the Lord's presence.

From habit, Simplicio regards the bell's toll as his personal alarm. He rises. His first act of the day mimics the habit of the monks. He initiates his daily devotional, opening his Bible to ingest the word of God. Today he reads from the Old Testament, from the book of Psalms. The text admonishes Simplicio to avoid the counsel of the ungodly, to abstain from the ways of the sinner, and to stand among the congregation of the righteous. Above all, it exhorts him to delight in the truth of the divine law of the Lord. [2] Each verse, imbibed solemnly and silently, permeates his being, instilling a sense of oneness with the Almighty. This ritual, executed every morning, day after day, year after year, not only comforts Simplicio's soul but fortifies his faith in the one true God.

The same ritual stirs moments of reflection, recalling to Simplicio the cardinal event of his life, one etched indelibly into his being: the moment God spoke to him. It still bewilders Simplicio that the Lord, in His eminence, singled him out as the unworthy recipient of a consequential message. Yet it happened. It was during prayer, at a moment when a singular calm had pervaded Simplicio's body, an unadulterated love had entered his heart, and a spiritual purity had penetrated his soul. At that moment, God spoke. Simplicio remembers the behest clearly, a mere three words: "Spread the gospel."

The recollection still causes Simplicio's chest to swell with pride. At times, so much so that he fleetingly embraces the seventh deadly sin, wondering whether he somehow earned the calling. A neutral observer might ridicule such a speculation but, to Simplicio, the prospect strikes him not at all as unlikely. In point of fact, he sometimes views the accolade as recompense for a life spent in preparation. The groundwork, you see, began in childhood, at the Methodist church one mile from his home, where he intoned the hymns, absorbed the sermons, and bowed his head in prayer. As a zealous congregant, he rarely was absent from a scheduled service. Speak to fellow penitents from the era and, most likely, they will recall the young boy, whose aspect, diction, and demeanor seemed to radiate spiritually.

At age twelve, Simplicio first stood at the pulpit. He had, on his own volition, memorized the *The Twenty-Five Articles of Religion*, the doctrinal liturgy of Methodism. Keen to impart the catechism, he recited the *Articles* to an awed congregation. Then, at age fifteen, he presented a monograph, one he himself penned, on Article 20, the oblation of Christ. The following year he authored a second tract, this time on Article 25, the Christian duty to obey civil authority. Both treatises confirmed, in the minds of the laity, Simplicio's reputation as a gifted, spiritual leader.

Similar feats followed in succession, securing Simplicio's position on the first rungs of the ecclesiastical hierarchy. Each triumph, it must be said, arose from a desire neither for honor nor acclaim, but from Simplicio's profound connection to the faith. Sensing the Lord's inspiration far more deeply than the average penitent, he unquestioningly accepted Jesus Christ as Son and Savior, affirmed the authority of scripture as conveyed in the Bible, and sought salvation through strength of faith. His spiritual beliefs not only dictated his sense of right and wrong, but guided his behaviors and shaped his worldview.

Who, then, can fault Simplicio, from the steadfastness of his conviction, for wondering whether he earned the summons from the Lord? Perhaps he did. He hopes so. But, to his credit, he has learned to temper such thoughts. He has learned to practice outward humility, subduing the sin of pride. And he has learned to rejoice in the calling itself, regardless of its cause.

It only remains to be said that Simplicio put the Lord's communication into practice. His first act was to enroll in a seminary, with the aim of becoming an ordained minister. While there, he poured over assorted Christian doctrines, examining each one with singular intent. His goal was simple. To handpick a denomination, the one best suited to spread the gospel. A notepad recorded his thoughts. Catholicism? No, too heathen. Presbyterianism? No, too intellectual. Anglicanism? No, too ecumenical, too compromising. Unitarianism? No, they foolishly disavow the Holy Trinity. Pentecostalism? Perhaps. But wait. Yes, here it is. Evangelicalism. The instant his eyes landed on the word, it resonated. It excited. It inspired. Simplicio's heart bore witness both to the truth of his find and to its ingress as a sign from God. Yes, he must retain his Methodist roots. But to pursue a more active dissemination of the gospel, he must become an Evangelical minister. And so he did.

We now return to Simplicio, just as he finishes his morning devotional. He ceases his recollections and the pleasant sentiments they invoke. A more pressing task awaits, one that will occupy his day and might extend well into the evening. He must prepare his Sunday sermon, a duty he will approach with ardent resolution. His wellspring of determination is threefold. First, he seeks to repay God's trust in calling him to the ministry, in selecting him as His emissary. Second, he receives inspiration from Holy Scripture, which affirms the importance of his work: "Go ye therefore, and teach all nations."[3] Last, and of arguably greatest import, he thirsts to extol the majesty of the Lord's presence, to impart the truth of His teachings, and to proclaim salvation through Jesus Christ the Savior. His congregation merits no less.

Simplicio will draft his composition in accordance with the conservative form of Evangelicalism that he has embraced. One that affirms the Almighty as the one and only God, whose tri-omni essence - omniscience, omnipotence, and omnipresence - attests both to His perfection and His authority. One that emphasizes the Holy Bible as His literal word. One that obliges the Lord's human creations to be "born again" as an attestation of their faith. One that promises salvation through acceptance of the Lord Jesus Christ, and through no other means. And one that favors vibrant spirituality over intellectual teaching. To Simplicio, these simple truths, eternal and unassailable, convey the presence of God to saint and sinner alike.

As Simplicio commits his thoughts to paper, he will adhere to a routine honed from repetition. He will draft his sermon, ponder it, revise it, pray about it, then rewrite it, over and over, three times, five times, seven times, until, to his mind, it conveys the resplendence of the Lord. Once he attains a workable draft, he will rehearse its delivery. He will stand facing a mirror. He will ensure his dress communicates confidence, his posture conveys assurance, and his facial expression imparts a holy countenance. When he speaks, he will begin softly, a practiced technique that he knows will quiet the assembly. He will pace his speech, implanting moments of silence after key exhortations, permitting the spiritual truths to penetrate. He will practice bowing his head in humility, raising his eyes to Heaven, and extending his arms outward in metaphorical embrace of the congregation, each gesture timed to an entreaty or admonition. As the spirit of the Lord fills his body, his voice will crescendo from piano to forte, its apex communicating the sanctity of the Lord's words. Yes, a full day's work lies ahead. The sermon and its rendition must be as perfect as Simplicio's skills allow.

Salviati also awakes, unwillingly, to the peal of the same bell. Eyes defiantly closed, he directs his first utterance to his smartphone. An impersonal, disembodied voice responds, announcing the time. Six o'clock. He swears. He remains horizontal, straining to return to sleep, but knows it unlikely. His mind wanders. His thoughts drift to the monastery, to its unwelcome bell, to its mere existence.

Salviati wonders what drives the inhabitants of the cloister, here in the twenty-first century, to so diligently sustain their belief in a Bronze Age mythology. He knows the rational answers, those provided by psychology, neuroscience, and evolutionary biology, among other sciences. He likewise knows religion negates the rational. Even so, the monks still amaze him. Their devotion to ancient fables, their delight in sophistic claims, their zeal for mysticism - he finds it baffling. And their bells disturb his sleep.

Just as Salviati drifts once more toward slumber, his alarm sounds. Seven-thirty. Startled awake, he curses a second time, then drags himself out of bed. He, too, must complete a task, one not dissimilar from that of his Evangelical neighbor. He, too, must set pen to paper. Salviati's manuscript, however, will not praise the Lord. It will not assert scriptural truth. It will not threaten eternal hellfire. It will, instead, extol the humble potato. Yes, the potato. More on that shortly.

Salviati, like Simplicio, was raised in Methodism. From early childhood, he sat in similar wooden pews, sang identical hymns, and endured the same tropes: exhortations to accept Jesus, entreaties to abjure sin, and interminable recitations of Biblical catchphrases. Yet, mostly, he daydreamed, anxiously awaiting liberation from each Sunday's torment. To Salviati, you see, the words rang hollow. He was quick to notice inconsistencies and fabrications: the circular, self-serving logic that "scripture is true, so find truth in scripture"; the touting of contradictory Biblical passages; the undisguised absence of objective evidence to substantiate alleged miracles; the assertion that faith alone reveals "the truth"; the incurious, sheep-like acceptance of ancient myths; and the proclamation of a caring deity who created his human subjects, yet sanctioned - even compelled - the most heinous crimes to destroy them.

The vacuous dogma of his religion stimulated Salviati to seek answers elsewhere. He began his search by examining the tenets of other faiths, hoping to discover a more coherent and intelligent theology. But here he encountered only disappointment. Without exception, he found the assertions of rival religions as fatuous and irrational as those of his own denomination. To his mind, each faith was as sophistic as the next.

Continuing his quest, Salviati next turned to history, and to various schools of philosophical thought. It was here, amidst much study and reflection, that he began to conceive an answer. Think of it, in its rawest form, as a mixture of empiricism and rationalism, doused with ample measures of science and ethics. Only later would he come to discern the mixture as secular humanism, a philosophy that constructs its worldview from reason and science. From the central precepts of the philosophy, Salviati came to discern its most salient trait: Humanism offered a different way of thinking, one that disregards faith-based assertions in favor of an evidence-based model founded on inductive reasoning.

He quickly apprehended both the uniqueness and potency of the model. No other philosophy, no other ideology, no other creed possessed anything similar. Its singular ability to distinguish likely fact from wishful fiction derived from a set of underlying components. To wit, from an emphasis on open inquiry as the crucial first step in acquiring knowledge. From the formulation of hypotheses, each predicated on testable predictions, and each prediction designed to permit alternative outcomes. From the demand for empirical data, diligently collected from well-designed experiments. From the dispassionate evaluation of evidence. From the insistence on independent verification of results. From a proficiency to detect error, so narrowing the venues where truth might be found. And, finally, from a respect for experimental outcomes, even when a confirmed or refuted hypothesis stands in stark contrast to tenaciously held beliefs.

Further study led Salviati to emulate the personal attributes that guide an evidence-based mindset. Among them, he focused on three: integrity, modesty, and flexibility. The first, integrity, demands an open, nonpartisan approach from the seeker of knowledge. He must exhibit a willingness to challenge existing paradigms. He must objectively frame inquiries, hypotheses, predictions, and experiments, resisting all temptations to coerce preferred outcomes. And he must champion a disinterested assessment of experimental results, free from ingrained beliefs.

The second attribute, modesty, displaces the advocate. It relegates the proponent of a hypothesis to an ancillary role. The expounder might exude intelligence and charisma. He might possess celebrity and authority. He might have garnered a litany of professional honors. Yet all are irrelevant to the confirmation or refutation of a hypothesis, which is judged solely from experimental results.

The third criterion, flexibility, challenges the notion of absolute truth. It posits all theories as provisional. It comprehends each theory not as a hunch, a speculation, or a conjecture, but as a well-substantiated explanation of phenomena informed from comprehensive bodies of evidence, yet one that might be revised or overturned by future discoveries. It affirms that no theory survives from deeply-held convictions. No theory persists from wishful thinking. No theory is set in stone.

The ethos of secular humanism led Salviati to another insight, one discovered by many before him but rendered no less remarkable by its rediscovery. It was this: The rise of a rational, evidence-based mindset marked the emergence of a new model, one rooted in the Enlightenment, aka the Age of Reason. The model oppugned religion's monopoly as the sole paradigm for explaining the world. It brought to an end the church's doctrinaire franchise, one held for millennia, to impose a faith-based archetype on all peoples everywhere. It annulled the imperative to ascribe poorly understood events to anagogic causes. It removed the obligation to impute mystery to an unseen, theistic entity. And in doing so, it wrought a profound evolution in thought, erecting a new paradigm that promised to explain - slowly, methodically, and credibly - the universe and all it contains, wholly free from the cudgel of cabalistic dogma. A paradigm that disentangled the existence of our species from supernatural authority, that unfettered humanity from the chains of obdurate thinking, and that displaced picayune compliance in favor of rational inquiry.

With these thoughts in mind, Salviati will approach his manuscript in a manner decidedly different from Simplicio. Salviati seeks no divine inspiration. He awaits no guidance from a deity.

He prays to no invisible entity. Instead, as a journalist, he investigates the topic. He seeks original rather than secondary sources. He interviews experts in the field, preferably those who have conducted recent research. Occasionally, he conducts his own research. From evidence, critical thought, and informed viewpoints, he will draft his piece, an article due to his publisher by weeks' end. This is the practice Salviati routinely follows, one that he now will apply to potatoes.

In this instance, he will examine the claim that the modest tuber, in part, enabled a smattering of European nations to impose their hegemony over a sizable portion of the world.[4] He will write about the starchy vegetable's first domestication some 7,000 to 10,000 years ago in South America. He will mention there are 3,000 to 5,000 varieties. He will chronicle its introduction to Europe in the sixteenth century, transported to the Old World by the Spaniards. He will affirm its ability to thrive in colder climates. He will document its nutritional yield as three times that of wheat per acre. He will explain how that same dietary boon led to population growth in Europe. He then will link the chain of increased population to societal sophistication, technological advancement, and world conquest. His job, of course, is to weave these various threads into a first draft, one that is both informative and engaging. Perhaps he will interject an amusing anecdote. It always helps.

Salviati enjoys researching and writing non-fiction. It aligns to his mindset. He sometimes envisions the entire universe as a giant jigsaw puzzle, one with trillions of interlocking pieces. Each confirmed discovery, each new scrap of knowledge, yields another piece of the puzzle. Sometimes the pieces fit easily; other times their edges preclude ready insertion. Salviati knows the puzzle will never reach completion. But gradually, carefully, over time, distinct images come to the fore, each revealing one of the universe's secrets. Today, Salviati hopes to insert the potato piece.

One might assume that Simplicio and Salviati eschew each others' company, as their respective worldviews conflict on an endless number of philosophical and practical topics: the existence of deities, the genesis of the universe, the age of the planet, evolution, global warming, gay marriage, and transgender rights - to cite but a few. Yet, perhaps surprisingly, the two ideologically diverse neighbors exhibit a none-too-common trait: they are friends. Their conversations permit them to hone their respective outlooks, to learn from each other, and to understand one another.

---

Miscellany

This and several other chapters close with assorted miscellany. Here are the first two instances.

In establishing Methodism, John Wesley rejected Calvin's doctrine of predestination, arguing instead that faith and good works mark the path to salvation.

Throughout the text, I interchange the adjectives *theistic* and *deistic* as synonyms.

{ Day One }

# On Absolutes

*Perhaps the greatest discovery in human history is that our traditional sources of belief are in fact generators of error and should be dismissed as grounds for knowledge.*[5]

Steven Pinker (1954-)

American Cognitive Psychologist

SIMPLICIO: Salviati, this is a surprise. How nice to see you.

SALVIATI: Good afternoon, Simplicio. It is a pleasure to see you as well. Do you come often to this café?

SIMPLICIO: Yes, I do, but normally at a later time. Will you join me for a coffee?

SALVIATI: Certainly. You seem in good spirits today.

SIMPLICIO: Indeed, I am. I just completed a counseling session with a member of my congregation. I am heartened by the conversation.

SALVIATI: It is kind of you to render aid to your congregant. I admire your willingness to assist those in need.

SIMPLICIO: Thank you, Salviati. I am but God's humble vehicle. I admit, however, it brings me joy when the wayward traveller regains the path of the Lord.

SALVIATI: Without breaching any confidences, may I ask the nature of the conversation?

SIMPLICIO: Certainly. The topic is not uncommon. The individual fell victim to temptation and sinned against God. He came to me with the desire to rejoin the righteous.

SALVIATI: Was his transgression notably grievous?

SIMPLICIO: Yes, I am afraid so. He sought the company of other men in a carnal, non-holy way.

SALVIATI: And your faith considers this a sin?

SIMPLICIO: I know you tease, but yes, it is a cardinal sin. Does Holy Scripture not affirm it immoral? Does the Lord not expressly declare it an abomination?[6] Many such passages proclaim it so.

SALVIATI: You quote a verse from the Old Testament, I believe. I have heard it remarked, though, that through Christ's sacrifice, the sin of homosexuality was replaced by the love and acceptance expressed in the New Testament. Is this not the case?

SIMPLICIO: Some might suggest the New Testament permits homosexuality, but no, such is not the case. Jesus's reference to "uncleanness" in Ephesians tells us so.[7]

SALVIATI: The word *uncleanness* can be interpreted in many ways. Perhaps it would be worthwhile to consult other, non-Biblical sources on the subject?

SIMPLICIO: No, there is no need to consult other sources. The Bible reflects the true word of God. It can be relied upon in every circumstance.

SALVIATI: Yet reliance on a solitary source, on this occasion a Biblical source, suggests an instance of circular logic, a declaration of *scripture is true, so seek truth from scripture*. Conclusions derived from this premise become self-verifying, as there is no independent benchmark to gauge their validity. Might scripture merely appeal to those who seek confirmation of their previously established and firmly held convictions?

SIMPLICIO: Your challenge, though honestly put, is misplaced. The Bible speaks truth. The disciples of Christ attested to its divinity, as do eminent theologians past and present. Untold laity likewise affirm its truths as the revealed word of the Lord. There can be no doubt.

SALVIATI: Yet, I wonder. Have you considered the verses might be in error? The scribes who authored the original text wrote in Hebrew, Aramaic, and Greek. Countless translations over past centuries reflect the knowledge and interpretation of sundry linguists and theologians. Human error - even bias - likely has altered the meaning of more than one verse. Perhaps the English word *abomination* alters the original intent of the first passage you cited?

SIMPLICIO: Your misunderstanding is familiar. Skeptics ignore the will of God. It is the Almighty who guides the hand of the translator, propelling an infallible rendering. Thus, the Bible, in all its translations, conveys the inerrant word of the Lord. It is He, Himself, who assures its accuracy. So yes, the transgression is doubtless an abomination.

SALVIATI: Yet your deity created our species, did he not? Why would he knowingly design a proportion, say, four percent of the population, with a proclivity to commit a proscribed sin?

SIMPLICIO: Yes, the book of Genesis attests to our existence as creations of the Lord. But to each, He allots unique challenges. None of us is perfect. It is the manner by which we address our imperfections that denotes our standing in the eyes of God. Some, from embracing the love of the Savior, resist their temptations and remain steadfast in their faith. Others, alas, succumb, risking not only their souls but their eternal salvation.

SALVIATI: To safeguard their souls, how do you determine what advice to give? In the case of this particular sin, how do you approach conversations with individuals who can benefit from your guidance?

SIMPLICIO: I follow the path prescribed by the Almighty, which is easy to discern. I start by asking the penitent whether he loves Jesus and accepts Him as his personal Savior. Healing can begin its course only if one accepts the love of the King of Kings. Happily, in this instance, my congregant answered in the affirmative. To instill confidence that he will be able to rejoin the righteous, I reassured him God loves homosexuals; it is homosexuality He abhors. I then explained no one is born gay; individuals turn to homosexuality from some deficiency in their upbringing. Perhaps he was in an abusive situation, perhaps he lacked a father figure, perhaps he lacked moral guidance, or perhaps he simply lived in a household closed to the spirit of the Lord.

SALVIATI: Did he offer insights as to cause?

SIMPLICIO: No, he was not comfortable disclosing that information. But together we built a plan, one that relies on counseling, church attendance, Bible readings, and, above all, prayer.

SALVIATI: So in this instance, you believe you have achieved your goal?

SIMPLICIO: Yes, I believe so. The penitent clearly felt a detachment from his Biblical gender. I explained change is possible, that one day he will revert to natural relations with women. He expressed, in turn, a craving to cleanse his soul of past sins, to once again embrace the spirit of the Lord, and to follow a path of repentance. I believe he will discover, through the grace of God, the strength to refrain from sexual sin. Yet you seem skeptical.

SALVIATI: Yes, I admit that I am. Your religion labels homosexuality a sin. The declaration is both unequivocal and uncompromising. There is no degree of gradation, no suggestion of tolerance. The stark, black-and-white doctrine is presented as an absolute, one of many such instances in your faith. Indeed, all faiths tend to express their assorted teachings as absolutes.

SIMPLICIO: I cannot speak for all faiths, but if the faith is true, then its teachings emanate from the one and only God. They cannot be denied.

SALVIATI: I note your words *if the faith is true*. Yet each faith believes it is the one true faith. Each believes its absolutes emanate from its deity, and from its deity alone. Each deems its absolutes to be true, exclusively so. One faith's assertions, of course, frequently contradict the absolutes of rival faiths. Yet each asserts deistic authority to justify its own absolutes, labeling the truths of other religions as false teachings.

SIMPLICIO: I concede this is the case. But the challenge of competing faiths is one of the tests God presents to us. Many embrace the truth of the Lord's teachings, while others stray into false beliefs. The Bible cites many such instances. Yet His actions toward the faithful who accept His word are clear. Did He not deliver the seven nations into the hands of the Israelites?

SALVIATI: I admit I am not familiar with the seven nations.

SIMPLICIO: The seven nations refer to the ancient enemies of the Israelites - the Hittites, Girgashites, Amorites, Canaanites, Perizzites, Hivites, and Jebusites. Scripture tells us all were mighty nations.[8] Yet their disbelief in the one true God precipitated their downfall; each was delivered into the hands of the devout. In this way, the Lord cast His blessings upon the faithful above all others. To our contemporary brethren, it might seem unclear why the Almighty permits so many false religions to exist, even to prosper. Yet we recognize it as a test of faith, a sign that we must remain steadfast in our beliefs. We know that, one day, He will reveal His full design to us. On that day, He will punish the wicked and reward the faithful.

SALVIATI: Your response clearly emanates from the heart, but I find such explanations suspicious. Have you noticed that interpretations of religious tenets invariably seem to coincide with the beliefs of the proponent? To the impartial eye, the interpretations seem contrived, solely to affirm an ingrained doctrine. Consider the host of inconsistencies in the Bible. Some passages, for example, cite that salvation is justified by faith alone.[9] Yet others expressly state that good works are required, that faith alone is insufficient.[10] The clever apologist, however, concocts a solution, one that forces a reconciliation of the contradictory passages and, in doing so, aligns to a preferred interpretation of deistic will.

SIMPLICIO: You fail to consider that what appears as an inconsistency to the skeptic presents no such challenge to the devout. Allow me to continue your example. I believe the passages you refer to appear in Romans and James. But they reflect no inconsistency. Their interpretation derives from context. In Romans, we stand before the Lord. It is through His eyes salvation is judged. Only He can see unerringly into our hearts. His judgment rests on faith alone. But in James, we stand before man, whose view is limited to the temporal world, and only imprecisely. It is man who renders judgment on man. Each of us justifies his brother from his works. So you see, there is no inconsistency.

SALVIATI: And yet, to me, such explanations smack of verbal trickery. They serve only to ratify a specific ideology. Permit me to extend the example further. Many passages, in a less enlightened age, were held to describe actual events. The world was created in six twenty-four-hour periods.[11] Jeremiah put on, took off, then hid a linen girdle as instructed by his deity.[12] An angel bound Satan in a pit for one thousand years. [13] Yet today, advances in scientific understanding lead many, perhaps most, to disbelieve such literal claims. To salvage the passages, ingenious minds reinterpret the verses, transforming them into allegories, each designed to reflect guidance from a deity, each attributed to divine authority. Their metamorphosis from actual event to handy apologue, if nothing else, belies their status as absolutes.

SIMPLICIO: I must disagree with you, Salviati. Holy Scripture always speaks the truth. Yet we mortals draw varied lessons at distinct times. We are imperfect beings, and the truths speak to us in ways commensurate to our needs. Regardless of interpretation, the Bible's underlying truths are immutable. Is it not written, "all scripture is given by inspiration of God"?[14]

SALVIATI: Yet consider the individual who favorably interprets the sacred writings of their own religion, but who quickly discerns irrational affirmations in the scriptures of a rival faith. How many Christians, for example, believe Muhammad flew to heaven on an animal "smaller than a mule and bigger than a donkey"?[15] Yet the same Christians find the notion of a talking serpent plausible.[16] This pattern pervades all faiths, each acting to sanction its own alleged truths, each willing to suspend the laws of nature for its own benefit.

SIMPLICIO: Perhaps so. But I reiterate, those who follow the one true faith are protected from the spurious claims of errant prophesiers. Indeed, the Lord explicitly warns us to "Beware of false prophets."[17]

SALVIATI: Yes, but consider further. Flying horses and talking serpents are clearly disprovable. Both animals lack the necessary body parts - wings and vocal cords - to perform their magical tasks. How have religions responded to such impossibilities? Have they disowned the existence of the fabricated creatures? No. All too often, each faith invokes deistic authority to affirm the reality of its preferred vertebrate. Indeed, for centuries, each has affirmed the absolute truth of their existence - the winged horse, the conversing serpent, or other such incarnations. No matter how firm the disproof, each undertakes every effort, employs every guise, to justify their authenticity. At the same time, each faith repudiates the equally preposterous assertions of rival religions. This technique, this deceit, permits each faith to maintain the pretense that its own beliefs possess no fault.

SIMPLICIO: Of course, I disagree. But does science not do the same? Science has its own absolutes.

SALVIATI: Some might suggest so, but they are mistaken. Science does not enshrine absolutes. It values theories, each founded on a comprehensive body of evidence, each substantiated by testable predictions, each explaining a broad range of phenomena, each correctly predicting the outcomes of new observations.

SIMPLICIO: Is the word *theory* not simply another name for an *absolute*?

SALVIATI: No, not at all. Theories, you see, come with a notable caveat. They always are provisional. Their truths are not immutable. New evidence might challenge the theory, causing it to be revised or discarded. This is a regular and critical aspect of science. Science is a self-correcting process, one that incessantly seeks new evidence to develop a superior understanding of phenomena. Indeed, the method of science stands in resolute contrast to that of religion.

SIMPLICIO: What do you mean?

SALVIATI: Religious truths derive from authority, the authority of a deity, revealed to the faithful through scripture, prayer, intuition, or other means. Science places little value on such first-person pronouncements. It is the formulation of testable hypotheses, impartially assessed from empirical evidence, that drives science's provisional truths. The result is a method unparalleled in its ability to confirm or refute professed claims. A method that never relies on the preeminence, prestige, or intellect of the claimant. A method that distrusts authority, both temporal and ethereal.

SIMPLICIO: Rest assured, Salviati; the Lord confers no first-person announcements. His truths are available to all who open their hearts to His glory. Did He not say, "no prophecy of the scripture is of any private interpretation"?[18]

SALVIATI: Yet history is replete with private, faith-based interpretations, is it not? If you will abide another example, take punishments. Floods, famines, pestilence, fire - in centuries past, all were interpreted as acts from a deity to punish the sins of mankind. This practice derived from a larger reality: For most of human history, there was no way to explain natural events other than as the mysterious acts of divine authority. No other paradigm existed.

SIMPLICIO: I do not, as yet, comprehend the direction of your example.

SALVIATI: Bear with me, please. Lightning rods illustrate the point. The lightning rod is thought to have been invented by the American polymath Benjamin Franklin (1706-1790). Its use spread across New England in the mid-eighteenth century as a means to protect physical structures and their inhabitants from lightning strikes. At roughly the same time, in 1755, an earthquake struck New England. Science drew (and still draws) no relation between the two events.

SIMPLICIO: That much seems reasonable.

SALVIATI: A creative Christian pastor named Thomas Prince (1687-1758), however, authored an inventive connection.[19] He saw lightning as a punishment from God, one the Almighty could direct at specific targets. According to Prince, the invention of the lightning rod challenged God's authority.

SIMPLICIO: In what way?

SALVIATI: The rods, by protecting individuals within their proximity, forced God to excise lightning from His arsenal of punishments. He could no longer direct a bolt at a distinct sinner.

SIMPLICIO: And the connection to earthquakes?

SALVIATI: Prince asserted that the Lord, bereft of electrostatic weaponry, invoked an alternate punishment, one far more devastating. Earthquakes. Here's the connection. Prince insisted that lightning rods direct electricity into the earth, charging the surrounding ground. The build-up of electrical charge, in turn, generates instabilities that invite earthquakes. From this explanation, he recast the earthquake of 1755 not as a natural event, but as an act of vengeance hurled upon the sinful from the mighty "Hand of God"[20]. The wicked, from their attempt to evade God's judgment through the invention and use of lightning rods, brought greater punishment onto themselves.

SIMPLICIO: You must admit, it is a creative connection.

SALVIATI: Yes, but to the neutral observer, the creativity of Prince's private revelation from the Lord is surpassed only by its absurdity.

SIMPLICIO: Your narrative suggests Thomas Prince was a false prophet.

SALVIATI: Perhaps. But few, if any, now believe earthquakes stem from the use of lightning rods. Prince simply concocted the fable to ratify a tendentious belief. He offered no evidence of any kind to substantiate his claim, electing instead to tout a faith-based explanation for a phenomenon poorly understood in his era.

SIMPLICIO: Then it would be incorrect to refer to Prince as a false prophet. He merely offered, as you suggest, an explanation for earthquakes long before an understanding of plate tectonics had emerged.

SALVIATI: Yes, a faith-based, evidence-free explanation for an act that, according to Prince, emanated from the absolute and incontrovertible will of his deity. Prince's act inspires a more general conjecture. Is it not reasonable to suggest that religious belief arose, countless millennia ago, as a way to interpret temporal phenomena when no other explanatory framework existed? Only recently has it become possible to reach more credible, fact-based explanations, free from supernatural incarnations. The catalyst, of course, was the emergence of the evidence-based method of science, a method that has disproved countless mystical claims. The disproofs, in turn, banished the myth of religious absolutes.

SIMPLICIO: You will forgive me, of course, if I disagree with you.

SALVIATI: Yes, I understand your reluctance to accept this conclusion. But consider a more serious problem that arises from absolutes. By declaring that a fabricated story or contrived explanation reflects the absolute truth of a deity, the quest to discover the actual truth becomes more onerous. When fable usurps reality, the desire to seek truth diminishes. A mythos takes hold in the mindset of the people, a mindset of misinformation that is transferred, generation to generation, perpetuating the misbeliefs. Those who embrace the mindset simply refuse to consider their fables might be mistaken.

SIMPLICIO: Yet you fail to explain why it is necessary to search more broadly when the truth already is known from the Lord.

SALVIATI: Your statement, perhaps unintentionally, confirms my point. Absolutes stifle inquiry. Religion instills a mental lassitude, a languorous mindset disinclined to seek answers outside its faith. The same mindset likewise discourages scrutiny within each faith. How many know, for example, of the merciless Christian sects in ancient Rome who, in their zeal to elevate a particular fabrication above the superstitions of rival factions, butchered not only frightened pagans but each other as well? How many know of the myriad church fathers who, disguised under a mantle of spirituality and infallibility, executed an endless litany of rapacious acts - simony, rapine, slavery, torture, murder, and still more. How many know of Christianity's method of advance, suppressing or destroying untold art, literature, sciences - and people - under the cleansing power of its faith? How many know of Simeon Stylites who, exhibiting the lunacy emblematic of countless monks and saints, constructed a platform atop a sixty-foot column in Antioch and resided there for thirty-seven years - frequently standing with arms outstretched in cross-like imitation - before expiring on his isolated, spiritual aerie?[21] How many Christians know these histories, or even want to know?

SIMPLICIO: You and I are unlikely to agree on these points. If one accepts Jesus Christ as Lord and Savior, if one opens their heart to His truths, if one is willing to repent of their sins, then all is revealed. The direction of one's life becomes clear. It is that simple.

SALVIATI: I suggest it is simpler still. People merely choose the book that is deemed sacred by the community into which they were born. From that book, and that book alone, they select the sections they like. It really is no more complicated than that.

SIMPLICIO: I suggest you underestimate the indisputable truths contained in the Bible.

SALVIATI: Perhaps I do. But if people were to dispassionately assess their own beliefs with the same scrutiny they apply to other faiths, they would attain the same conclusion many already have reached: Their allegedly sacred texts were crafted by ordinary people, not by deities. As such, the texts simply reflect the oral traditions common to their era. Nothing more.

SIMPLICIO: Yes, crafted by ordinary people, but whose hands and hearts were moved by God to spread His word.

SALVIATI: Touché, my friend. Would you like another coffee?

SIMPLICIO: I'm afraid not. I have to be off now. I have to put some finishing touches on my sermon in preparation for tomorrow's services. I enjoyed our conversation though. Shall we meet at the same time tomorrow afternoon?

SALVIATI: Yes, let's. Goodbye, Simplicio.

SIMPLICIO: Blessings, Salviati.

SALVIATI: (to himself, while walking away) Alas, poor Simplicio. His absolutes circumscribe his perspective.

SIMPLICIO: (to himself) Alas, poor Salviati, his heart is not open to the way of the Lord.

*End of the First Day*

A dozen benefits of religious absolutes. **Diagram 1.1**

### Religious Absolutes

My belief in absolutes permits me to:

1. Attest my absolutes are true, as they derive from a deity.
2. Confirm my faith, and my faith alone, embodies absolute truth.
3. Discern the irrationality of other faiths, but affirm the rationality of my own.
4. Dismiss the need to question the tenets and history of my faith.
5. Subdue the temptation to seek answers outside my faith.
6. Sustain my absolutes even when contradicted by facts.
7. Contrive inventive interpretations of conflicting scriptures.
8. Indoctrinate my offspring.
9. Mistreat those who reject my absolutes.
10. Dismiss, even purge, uncomfortable facts and opinions.
11. Refuse my absolutes might be mistaken.
12. Eliminate the need to think.

{ Day Two }

# On Misuse Of Absolutes

*Rational arguments don't usually work on religious people. Otherwise there would be no religious people.*[22]

House

SALVIATI: Hello, Simplicio.

SIMPLICIO: Good afternoon, Salviati. I hope I'm on time.

SALVIATI: Perfectly so. Yet you seem out of breath.

SIMPLICIO: Yes, I walked hurriedly, nearly sprinted, directly from the church. I did not want to be late.

SALVIATI: Rest assured, you are on time. And even if you were late, you needn't worry. I always carry a book to read; it helps to fill unexpected, quiet moments during the day. I recall you were delivering your sermon this morning. Did it go well?

SIMPLICIO: Yes, I am pleased. I instructed my flock on the perils of Hell. It is my duty, of course, to safeguard their souls.

SALVIATI: Indeed. Were you able to convince your parishioners of the existence of Hell?

SIMPLICIO: Why, of course, Salviati. There can be no doubt.

SALVIATI: How do you know Hell exists?

SIMPLICIO: Does not Jesus explicitly acknowledge the existence of Hell?[23] It would be imprudent to ignore His word.

SALVIATI: Perhaps so. Is it also true Hell is an unpleasant place?

SIMPLICIO: I know you bait, but yes, it most certainly is an unpleasant place. It is the abode of Satan and his demons. It likewise is a place of punishment for those who reject Christ. [24] Does the Lord not declare it a den of everlasting fire?[25]

SALVIATI: Perhaps so. But once consigned to Hell, can a sinner obtain redemption and escape his eternal fate?

SIMPLICIO: No, Salviati. Once dispatched to the lake of fire, the sinner must spend an eternity in torment and destruction. Such is the will of the Lord.[26]

SALVIATI: Where is Hell located?

SIMPLICIO: It is written Hell lies within the earth.[27] There is a chasm within the planet, a large chamber where the sinners reside.[28]

SALVIATI: You are aware, of course, how the earth formed? It condensed from the primordial nebula, the cloud of dust and gas that pervaded our region of the cosmos. Our home star and its planets formed from the gravitational attraction of the physical granules. It is not clear how this process could carve out a bespoke chamber within the earth.

SIMPLICIO: Yet scripture affirms the chamber exists. Perhaps it resides beyond sight in the center of the earth.

SALVIATI: But we know the earth's core is solid. During the earth's formation, the more massive elements, the metals, sank to the core under the force of gravity. The lighter elements, those that comprise rocks, remained nearer the surface. As the earth cooled, the core began to crystallize.

SIMPLICIO: But we do not know this. We are unable to penetrate the mysteries at the core of our planet.

SALVIATI: Actually, we do know this. Seismic events - earthquakes, volcanic eruptions, massive landslides - send sound waves through the planet. We know that one type of wave, primary waves, passes through any material. A second type, shear waves, passes only through solid objects. By analyzing the speed, frequency, and direction of travel of the different types of waves, we can ascertain the density and elasticity of the materials in the various layers of the earth. And we can determine the core is solid. Solid iron.

SIMPLICIO: Yet is it not but a theory?

SALVIATI: Yes, it is a theory, a theory constructed on evidence. Findings in other fields support the theory - measurements of gravitational and magnetic fields, and analyses of objects thrown up from volcanos.

SIMPLICIO: But your so-called theory does not shed light on the origin of the burning fires of Hell, is that not so?

SALVIATI: Burning fires, no. But high temperatures, yes. We know that temperature, like pressure, increases with depth. We also know the source of the lofty temperatures within the earth. They arise from the initial gravitational friction as the planet formed, from the friction of denser materials sinking to the core, and from the decay of radioactive elements.

SIMPLICIO: If temperature intensifies with depth, the core indeed must be the location of Hell, as it is the hottest place within the earth.

SALVIATI: Yes, it is the hottest location. Its temperature is estimated to be around 6,000 degrees Celsius, akin to the surface temperature of the sun, perhaps even hotter. And the pressure in the core is three million times greater than on the surface of the earth.[29] No organism could survive in such conditions.

SIMPLICIO: Then perhaps Hell resides elsewhere under the surface of the earth. Scripture also refers to Hell as a pit.[30] Perhaps the pit lies closer to the surface of the planet.

SALVIATI: Perhaps, but seismic measurements should be able to locate a vast chamber or pit. One could conduct an experiment. Place measuring devices - seismometers, hydrophones, accelerometers - in various locations. Then monitor sound waves from natural events, or even from purposely-set explosions. Over time, one could map the likely locations for subterranean cavities.

SIMPLICIO: But what if your so-called experiment fails to discover the location of Hell?

SALVIATI: If a sufficient number of experiments demonstrate there is no Hell, you can allay the anxiety of your followers. You can assure them Hell is an illusion. They will never face an eternity of punishment. Think of the joy they will experience.

SIMPLICIO: But this cannot be. The Lord commands the existence of Hell. He has confirmed many a soul has been sent to Hell.[31] We cannot disavow His teachings. If your approach disproved the existence of Hell, there could be only one explanation. It would be a sign that God wishes to test our faith. We must accept the word of the Lord over temporal claims.

SALVIATI: If that is the case, allow me another inquiry. Have you thought how sinners find their way to Hell?

SIMPLICIO: What do you mean, Salviati?

SALVIATI: There is a wealth of questions one might ask. What is the soul? Where does it reside within the physical body? How does it separate from the body? How does it determine the location of Hell? How does it navigate to the site? How does it travel through a solid earth to reach the chamber? How does it survive the high temperatures and pressures? There are myriad such questions.

SIMPLICIO: Yes, there are innumerable questions. But it is not our way to question the acts of the Lord. He reveals His imperishable truths to us at a time and place of His choosing. Does scripture not pointedly counsel us to avoid "profane and vain babblings, and oppositions of science falsely so called"?[32]

SALVIATI: Have you considered scripture might be mistaken on this point?

SIMPLICIO: That is not possible. Scripture imparts incontestable truth. Scripture is the word of the Lord.[33]

SALVIATI: Yet there seems to be a consistent dearth of evidence regarding all facets of Hell. Would you not welcome more rigorous inquiry, inquiry that might support your claims?

SIMPLICIO: It serves no purpose, Salviati. Only the unbeliever questions the will of the Lord. The overly inquisitive individual rarely finds respite in the House of God.

SALVIATI: I understand the dilemma, Simplicio. I would welcome further discussion, but I'm afraid it is I who must bring our conversation to a close today. I am working on a piece that soon is due to my publisher. Several hours of work still lie ahead.

SIMPLICIO: I, too, must rush. I fear I am late preparing for a Bible study session. Same time tomorrow?

SALVIATI: Yes, I look forward to it.

SIMPLICIO: Allow me to wish you blessings, Salviati.

SALVIATI: Until tomorrow, Simplicio.

SIMPLICIO: (to himself, as he departs) To call into question the existence of Hell is a sign of Salviati's theological innocence.

SALVIATI: (to himself) The firmness of Simplicio's convictions permits little light to enter.

*End of the Second Day*

{ Day Three }

# On Evidence

*The reasonableness of something is rarely a motive in human affairs.*[34]

Barbara Tuchman (1912-1989)
American Historian

SALVIATI: Hello, Simplicio. From the measure of coffee in your cup, it appears you arrived early today.

SIMPLICIO: Yes, I did. I wanted to ponder the nature of our last two conversations. I admit I was troubled.

SALVIATI: I hope I left you with no conundrum. May I ask what concerns you?

SIMPLICIO: I trust you will take no offense, but it is clear you suffer from weakness of faith. Your words suggest you do not extend adequate consideration to the existence of our Lord and His teachings. Perhaps you would allow me to ask a question, if for no other reason than to better understand your thinking.

SALVIATI: Certainly, Simplicio. Pose any question you wish.

SIMPLICIO: Please tell me, what would it take to convince you of the truth of our faith? What would open your eyes and heart to Holy Scripture? What would persuade you to at least consider the validity of our Savior's words?

SALVIATI: You raise legitimate questions, Simplicio, ones that warrant a thoughtful response. I think you know I am open to believing anything, subject to a single prerequisite. It must be founded on testable predictions. To understand this, permit me to respond with a recipe of four ingredients.

SIMPLICIO: By all means, I am anxious to hear your thoughts.

SALVIATI: I start with what I hope you concur is a sound, rational principle. The onus is on the claimant to demonstrate the validity of a claim. It is not the responsibility of the skeptic to disprove a claim, even though she might be able to do so. This principle applies equally to science, religion, and all other disciplines.

SIMPLICIO: That sounds reasonable.

SALVIATI: From that cornerstone, let me proceed. I refer to the first of the four ingredients as *preponderance*. To explain it, consider Einstein's special and general theories of relativity, which introduced a new paradigm of gravitation, a model that departed sharply from previous physics. To paraphrase Sagan's dictum, it was an extraordinary claim requiring extraordinary evidence.[35] Einstein and others satisfied the stipulation of extraordinary evidence by formulating a number of detailed, testable predictions.

SIMPLICIO: Can you remind me of the predictions?

SALVIATI: Certainly. Allow me to enumerate several of them.

- The equivalence of mass and energy ($e=mc^2$).
- The deflection of light in the presence of massive objects.
- The related phenomenon of gravitational lensing, created when a massive body intervenes between a light source and an observer.
- The reduction in the frequency of electromagnetic radiation in response to gravitational fields (gravitational red-shifting).
- The presence of intensely massive objects (black holes).

- The existence of frame dragging, the twisting of spacetime around a spinning, massive object.
- The constancy of the speed of light, at 299,792,458 kilometers per second, regardless of the speed of travel of the observer.
- The occurrence of gravitational waves, ripples traveling through spacetime at the speed of light, driven from the acceleration of massive objects (neutron stars, black holes).
- The assertion a clock runs more slowly on the ground floor of a tall building than on the top floor.
- The prediction a clock in motion runs more slowly than a clock at rest.
- The affirmation a rod in motion will be shorter than a rod at rest.
- The conjecture you will age more slowly if traveling in a spaceship at high speed.
- And still more.

SIMPLICIO: It is a lengthy list.

SALVIATI: Yes, but of greater consequence, each prediction was testable. Each suggested a well-defined outcome which could be confirmed or disproved - unequivocally. And I believe you know the result.

SIMPLICIO: Yes, I do.

SALVIATI: To date, a diverse set of experiments performed by a disparate assemblage of independent actors have confirmed Einstein's predictions, all to a precise degree of measurement. Some of these actors even designed their experiments with the specific intent to disprove, what they regarded, as Einstein's "unsavory" theories. Yet they, too, confirmed his predictions.

SIMPLICIO: That is an impressive outcome.

SALVIATI: Yes, it is. And let me emphasize the words *to date*, as they carry further significance. Tests of Einstein's theories began in the early twentieth century and continue to this day: new experiments exploiting new techniques, yielding ever higher degrees of precision. And so it will continue. In the life of an extraordinary claim, the moment of surrender never arrives. There is never a point of acquiescence, never a proclamation of *we've adequately verified the predictions, that's that, let's accept the result, let's launch no further inquiry, let's rest on our laurels*. The verification of an extraordinary claim is continual: unceasing inquiry, unremitting challenge, unending re-verification.

SIMPLICIO: This is why theories are not accepted as absolute truth?

SALVIATI: Yes, that is one reason. But more generally, as I hope my example underscores, there is no single event, no single prediction, no single moment in time that can substantiate an extraordinary claim. The bar is higher. A broad and robust range of detailed predictions is required. If borne out by subsequent experiment, the test of preponderance is temporarily satisfied, but always is subject to reassessment at later dates.

SIMPLICIO: You are suggesting religious claims meet the same bar.

SALVIATI: Yes, I am. Religious claims of a supernatural being are several magnitudes more extraordinary than Einstein's assertion of a new gravitational model. Religion thus requires a correspondingly greater number of ironclad, testable predictions. The onus is on religion to formulate its predictions, attended by a rigorous, transparent testing framework. This is the ingredient of *preponderance*.

SIMPLICIO: What is the second ingredient?

SALVIATI: It is *novelty*. To describe it, I will continue with the example from Einstein. His theories illuminated previously unexplainable phenomena. They generated predictions that few people, if any, had previously envisaged. They were novel.

And, when first formulated, no technology existed to test many of the conjectures. This means no one could have furtively conducted a test, confirmed an outcome, crafted a theory aligned to the outcome, published the theory, then touted an after-the-fact prediction. Religion likewise must embrace novelty. Among its testable predictions, it must describe novel events and foretell pioneering outcomes that cannot be fudged.

SIMPLICIO: Would you consider the Second Coming of Christ to be novel?

SALVIATI: It certainly would be a novel event, but insufficient in context of a novel prediction. In the last two millennia, countless prognostications foretold the second coming. All were disproved, without exception. Continuing to predict the same event, over and over, despite repeated failures, negates the credibility of both the claim and the claimant. To meet the test of novelty, the prediction must be specific enough to denote time, place and manner. And, as with preponderance, it must incorporate a testing framework.

SIMPLICIO: What type of testing framework?

SALVIATI: If an entity arrived on our world, claiming to be a supernatural being, how would you assess the truth of its claim? How would you differentiate between a supernatural being and an inhabitant from an advanced planet, one who could mimic supernatural powers? How would you distinguish between a supernatural entity and a deity, as the two are not the same?

SIMPLICIO: Could we not trust our hearts to interpret the truth of the entity's words?

SALVIATI: As an adherent of the scientific method, I answer no. A novel prediction requires an objective framework, one established prior to the arrival of the entity, one that specifies clear conditions for accepting or refuting the claims of the entity. Perhaps I can portray the importance of a predefined framework with the following. Suppose an entity arrived, claiming to be a prophet. The entity proclaims the truth of a rival religion, denouncing your faith as false. Would you accept the decree?

SIMPLICIO: I admit it would be difficult.

SALVIATI: Any religion would find it so. This reality evinces the importance of a pre-established and robust testing framework. It limits post-event reinterpretations and other types of wiggle room.

SIMPLICIO: It is a fair point.

SALVIATI: Let's move on to the third ingredient, the one that is perhaps most difficult to explain. I refer to it as *method*. Two days ago, I alluded to the fact that science derives its truths from the scientific method, from a way of thinking that demands critical inquiry, the formulation of testable hypotheses, the establishment of robust testing frameworks, the collection of empirical data, and the unbiased analysis of results. In competent and nonpartisan hands, all steps lead to the confirmation or refutation of well-constructed hypotheses.

SIMPLICIO: Yes, I am aware of this.

SALVIATI: Religion reaches its truths via a different method, a faith-based method that accredits divine pronouncements, sacred writings, and privy attestations, all derived from theistic authority. To demonstrate the legitimacy of its philosophy, religion must address preponderance and novelty in like manner, formulating its predictions from its faith-based method of thinking. It cannot simply borrow techniques from the scientific method. The reason should be obvious. If religion can formulate testable predictions only from embracing the scientific method, it is an admission its own faith-based method is both impotent and irrelevant.

SIMPLICIO: I will not say I agree with you, but please continue with your argument.

SALVIATI: How might religion formulate testable predictions from its faith-based method of thinking? To answer this somewhat rhetorical question, allow me to refer to the Pythagorean theorem, which was discovered some three to four millennia ago in Babylonia. You will recall that the theorem expresses (among other things) the lengths of the sides of a right triangle: $a^2 + b^2 = c^2$, where each letter represents one side of the triangle. There is an infinite number of positive integer solutions to the formula. Let's refer to such solutions as triples. One triple is $a = 3$, $b = 4$ and $c = 5$. Other triples include (12, 35, 37), (17, 144, 145), and (37, 684, 685).

Pythagorean Theorem

Diagram 3.1 Finding a triple solution for a right triangle, where $a^2 + b^2 = c^2$.

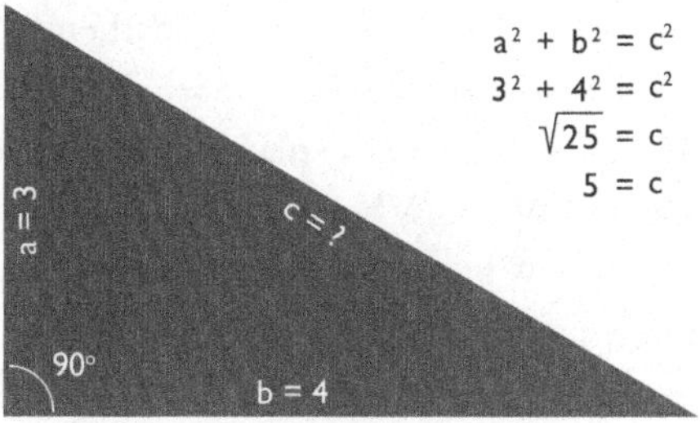


SIMPLICIO: Yes, I recall the theorem from school.

SALVIATI: You might also recall that, in the seventeenth century, the French mathematician Pierre de Fermat (1601-1665) explored a more general form of the equation: $a^n + b^n = c^n$. He asserted that, when *n* is greater than 2, there are no triple solutions.[36] Fermat hinted he had developed a mathematical proof of his assertion but, upon his death, no such documentation was found among his papers. Over the next 329 years, a progression of clever and accomplished mathematicians attempted to devise a proof. The British mathematician, Andrew Wiles (1953-), at last succeeded in 1994, affirming Fermat's assertion.

SIMPLICIO: Interesting, but how does Fermat's assertion fit into our conversation?

SALVIATI: Imagine the following. An archaeological dig uncovers primeval pottery, ten thousand years old. Inside an unopened pot rests an antediluvian scroll, a parchment. Luminescence dating corroborates the antiquity of the pot. Carbon dating verifies the age of the parchment and its ink. Linguistic analysis corroborates the authenticity of the alphabet, vocabulary, syntax, and historical context of the ancient script. Independent translations reveal the content of the scroll: a proof of Fermat's assertion. A robust, mathematical proof, inscribed well before the discovery of the Pythagorean theorem by any known society on the planet, and long before the birth of Fermat and Wiles. What prominence might our imaginary scroll assume? What claims might religion assert? Religion could interpret the scroll as an expression of divine knowledge, revealed from a creator at a time when no member of *Homo sapiens* possessed such knowledge - a claim that would be in accordance with religion's faith-based method of thinking. Religion could further assert that the authenticity of the scroll and its contents were verified independently by science, via radiocarbon dating and other methods.

SIMPLICIO: That would indeed be a unique claim. Religion then could argue that science cannot dismiss the scroll simply as a record of the oral myths of a parochial people.

SALVIATI: Exactly. Now let's extend the example. Imagine a robust catalog of similarly detailed, testable discoveries, derived from a cache of newly-uncovered primeval scrolls, from revelations disclosed to contemporary prophets, or from fresh scriptural interpretations: unambiguous cures for as-yet unidentified diseases; a formula unifying quantum mechanics and gravity; the exact planetary coordinates of an intelligent, communicating species of extraterrestrial life; and, from the theological realm, an explicit and detailed rendering of the mechanism by which souls ascend to heaven. You grasp the idea. If religion could produce such a catalog, it would be difficult for science to summarily discount the claims.

SIMPLICIO: I would welcome such a catalog.

SALVIATI: How might a cache of revelatory scrolls be unearthed? Must we await chance discovery by archaeologists? No, we need not. Religion can, here again, demonstrate the efficacy of its faith-based method of thinking, this time by way of prayer. Employ the presumed agency of prayer to persuade a deity to reveal the location of the scrolls - their longitude, latitude, and depth. Humbly beseech the creator to disclose their locus to a contemporary prophet. If the scrolls do not yet exist, entreat the deity to fashion extra-temporal scrolls, a simple feat for an allegedly omnipotent deity. Imagine the result. Parchments that not only contain divine knowledge, but also substantiate religion's claim for the efficacy of prayer.

SIMPLICIO: I follow your reasoning, but I wonder whether the Lord would welcome supplications for such precise guidance. He might interpret the entreaties as a sign of unbelief. Perhaps the faithful better serve the Lord by awaiting His revelations.

SALVIATI: Perhaps, but allow me a speculation. It is odd that no revelation - not one - from a purportedly omniscient deity demystifies an unexplained natural phenomenon. No scripture discloses the testable tenets of an original, unheard-of, earth-shaking field of knowledge. No sacred text contains a single graph, equation, or calculation that conveys a robust, mathematical comprehension of phenomena. Why not? The most plausible explanation is the self-evident one. Revelation and scripture arise in no respect from divine origins, but merely from the fertile minds of *H. sapiens*. As such, the writings deciphered from various parchments, papyri, and palimpsests reflect nothing but the contemporary knowledge of their sectarian authors.

SIMPLICIO: I cannot accept this conclusion. The ways of the Lord often lie beyond our understanding.

SALVIATI: Yes, I know it is difficult. Allow me, however, to proceed to the fourth and final ingredient, *integrity*. Let's presuppose religion furnishes the first three ingredients. It produces a voluminous catalog of novel discoveries from its faith-based thinking process, replete with detailed explanations and testable predictions. Now what? If the predictions are confirmed, the onus shifts to science to consider the religious claims. Science should acknowledge that the ingredients form a rich broth, disclosing a valid theological framework for explaining the world we inhabit. But the corollary is equally pertinent. What if the predictions are disproved?

SIMPLICIO: I sense where you are going.

SALVIATI: I know you do, but humor me a moment longer. Let's briefly return to the special and general theories of relativity. What if experiments failed to corroborate Einstein's predictions? What if his predictions were refuted? What if a preponderance of evidence favored a radically different model? Then science would concede Einstein's theories are wrong. Period. As Feynman expressed, the fact that a prominent individual formulated the theories is meaningless.[37] The brilliance, reputation, and persuasive powers of the proponent

(Einstein, in this case) are inconsequential. Integrity requires that religion play from the same rulebook. If experiments disprove faith-based claims, religion must acknowledge the findings, cease its claims, and reject its dogma. This is daunting, as religion on every occasion invents an "out" to alter, discredit, reinterpret, or otherwise dismiss unwelcome evidence. Of the four ingredients, the challenge of integrity is the most exacting.

SIMPLICIO: Your recipe is clear. Perhaps, one day, we will be able to deliver all its ingredients.

SALVIATI: Perhaps, but my guess is that religion will be unable to produce a catalog of novel, testable discoveries derived from its faith-based thinking method. Yet, to repeat an earlier comment, I will believe anything if supported by objective evidence. I am open to new paradigms and, at least in spirit, I await your catalog.

SIMPLICIO: Thank you for responding to my question so candidly and thoroughly. You have given me food for thought.

SALVIATI: Allow me to thank you as well. Your willingness to entertain an alternate view is refreshing. I look forward to continuing our discourse tomorrow.

SIMPLICIO: I, too. Blessings, Salviati.

SALVIATI: (to himself, as he walks home) Simplicio is a kind soul. Perhaps his faith does not preclude all reason.

SIMPLICIO: (to himself) How I wish Evangelicalism possessed the ingredients of Salviati's recipe, to precisely and unequivocally demonstrate the proof of God's words.

*End of the Third Day*

---

Miscellany

The American physicist and Nobel Prize winner, Robert Milliken (1868-1953), reportedly expended considerable energy attempting to disprove Einstein's particle theory of light, which Milliken believed was incorrect. In 1915, after ten years designing and conducting experiments to disprove the theory, he publicly (and perhaps reluctantly) confirmed Einstein's predictions.

The origin of the Pythagorean theorem is disputed. Some suggest it first was discovered in Babylon. Other scholars point to Egypt, India, China, and elsewhere, where a more complete understanding of the theorem arguably emerged. Early Greek and Roman literature attribute the discovery to Pythagoras (c.570-c.495 BCE), a Greek philosopher and mathematician, as the eponymous source of the theorem.

**Diagram 3.2** Four mandatory ingredients to substantiate religious claims.

### A Recipe for Religious Evidence

| | | |
|---|---|---|
| ① | Preponderance | Predict a broad range of phenomena. |
| ② | Novelty | Predict novel phenomena. |
| ③ | Method | Derive predictions from religion's method of faith-based thinking. |
| ④ | Integrity | Accept the implications of refuted predictions. |

{ Day Four }

# On Absence Of Evidence

*Man is a rational animal - so at least I have been told.*[38]
Bertrand Russell (1872-1970)
British Philosopher

SALVIATI: Good afternoon, Simplicio. Have you reflected further on our previous talk?

SIMPLICIO: Indeed, I have. Last night I contemplated the evidence-based recipe you proposed, running through its ingredients as sleep approached. While I respect the clarity of the recipe, as a servant of the Lord I continue to believe some mysteries are discernible from faith alone. Is it not true that science, despite its evidence-based methods, cannot explain everything?

SALVIATI: It most certainly is true. There are many things science cannot yet explain. But how one responds to a lack of explanation is, of itself, informative.

SIMPLICIO: In what way is it informative?

SALVIATI: Allow me to cite an example. Consider bird migration. Many aspects of migration are still incompletely understood.

SIMPLICIO: Yes, I am aware of this fact, but not of the particulars. What remains to be discovered?

SALVIATI: Much remains. Researchers seek answers to sundry

open questions. Why do some species migrate, others not? Why do some individuals within a species migrate, others not? Why are younger birds more likely to migrate than older birds? What physiological changes allow the northern wheatear, an insubstantial creature weighing the equivalent of twenty paperclips, to migrate over 7,000 kilometers from Eastern Canada to Sub-Sahara Africa? How do birds know when it is time to migrate? Is migration a response to variations in weather, changes in daylight, or obedience to an internal clock? How do birds choose their destinations? How do they navigate to their destinations? Do they pilot from the position of the sun, from the strength of the earth's magnetic field, or from olfactory cues? How do they navigate back to their points of origin? How do they navigate across oceans possessing no unique landscapes? How much of their navigational skillset is learned versus inherited? Why do some birds migrate in groups, others individually? Why do some birds migrate in daylight, others at night?[39]

SIMPLICIO: Much of this knowledge seems fundamental. I am surprised so many open questions remain. But again, in what way is this lack of knowledge informative?

SALVIATI: When knowledge and understanding are incomplete, one can respond in one of two general ways. The practitioner of evidence-based thinking attempts to understand the phenomenon. She will pose thoughtful inquiries. She will assess pertinent information. She will challenge accepted assumptions. She will weigh contradictory evidence. She will identify patterns beyond immediate context. She will isolate relevant variables. She will diagnose faulty logic. In the end, she will reach a conclusion from confirmed and refuted predictions, each disclosed from experiment. This approach, repeated by independent researchers, gradually discloses the knowledge.

SIMPLICIO: Yes, I recall your scientific method, and I am aware of its frequent sluggishness. But you mentioned it as one of two general approaches?

SALVIATI: Indeed, I did. If I do not believe in the scientific method, if I prefer not to wait for its objective results, if I crave another, more immediate paradigm, I can invent a deity. In this instance, an avian deity. One that guides its flocks from a celestial aerie. An omniscient, omnipotent, and omnipresent bird-deity that directs the thoughts and deeds of its avian subjects. I can ascribe to it qualities of compassion, caring, and love for its winged creations. I can name it Ornith.

SIMPLICIO: Salviati, I shall never tire of your humor.

SALVIATI: But envisage the benefits. Any inquiry concerning bird migration can be simply satisfied. Why do birds migrate? It is the will of Ornith. How do birds locate their migratory destinations? The bird-lord reveals the necessary knowledge to his volant progeny at a time of his choosing. Why do some birds migrate in daylight, others at night? We are not meant to comprehend the will of the ornithoid lord. It is all so simple, is it not?

SIMPLICIO: Your implied comparison is indeed entertaining. But it would be blasphemous to suggest the Lord is an invention. His very Being always has existed and ever will. There is no need of invention.

SALVIATI: But consider further. If anyone challenges our tale, we will admonish the doubter to lay open his heart to Ornith. We will inspirit the cynic to engage in prayer, to trust in the truth of his feelings. But we must take caution in our fable to select only those "proofs" that cannot be tested, cannot be empirically examined, cannot be disproved. We must design our fabrication to be irrefutable.

SIMPLICIO: Ah, Salviati. You reveal your bias. I know you value objective evidence over the revealed truths of our Lord. Have you now concluded your monologue on the mythical bird-faith?

SALVIATI: Allow me one more proposal. To project an aura of ethereal origin, we can cloak our fiction in an alloy of medieval and mystical prose. And we can add the threat of punishment for any who fail to embrace our fable. Perhaps something like this: *And lo, thou hearest now the word of Ornith, He who hath bestowed upon His apostles the spirit of life, He unto them. Thou shalt rejoice, thou and thine household, before the winged Almighty, and receive His blessings. But beware, lest thou forget the teachings of the avian Lord and provoke His wrath. Unto thee His disciple He sayeth, verily, profane not the apocrypha, as thou fearest eternal hell-fire.*

SIMPLICIO: Enough, Salviati. I commend your attempt at parody. You have amused us both. I know you merely wish to point out that a paucity of knowledge does not justify the creation of a deity.

SALVIATI: Yes, that is correct. The invention of a deity, the attribution of unexplained events to an invisible being, an avowal of mystical truths - absence of knowledge justifies none of these.

SIMPLICIO: Your point is both clear and diverting. But now I must beg your pardon. The hour is late and the hottest part of the day has passed. I will smile at your musings on the walk home.

SALVIATI: Until tomorrow, Simplicio.

SIMPLICIO: Blessings, Salviati.

SALVIATI: (to himself) Simplicio is kind to tolerate my banter.

SIMPLICIO: (to himself) Perhaps I should disclose to Salviati that I concede his point, in part. There is no doubt the absence of knowledge led to the false gods and profane writings of non-Christian faiths. Christianity alone discerns the one true God.

*End of the Fourth Day*

Faith-based and evidence-based approaches to knowledge. **Diagram 4.1**

| Religion | Reason |
|---|---|
| Phenomena exist that I do not understand, so I: | |
| ① Invent a deity. | ① Acknowledge the unknown. |
| ② Ascribe the phenomena to the will of the deity. | ② Conduct research. |
| ③ Declare immutable truth. | ③ Establish theories based on testable predictions. |
| ④ Seek no other explanations. | ④ Revise as new discoveries warrant. |

{ Day Five }

# On Labels

*To you I'm an atheist; to God I'm the loyal opposition.*[40]

Woody Allen (1935-)
American Filmmaker

SALVIATI: Good afternoon, Simplicio. I must admit, I look forward to our daily conversations.

SIMPLICIO: I, too, Salviati. Shall we continue our discussion of evidence, both its presence and its absence?

SALVIATI: If you don't mind, I thought we might take the conversation in a different direction.

SIMPLICIO: Certainly, Salviati. Allow me to say, however, you appear pensive today.

SALVIATI: Yes, I was thinking how people are labeled by their beliefs. The label *Christian* is applied to followers of Christianity; *Muslim* to Islam; and *Jew* to Judaism. Each label carries various denotations and connotations, but all allude to some higher faith, some hallowed realm. Yet the word *atheism* seems to elicit a different treatment, one I struggle with.

SIMPLICIO: Why is that, Salviati?

SALVIATI: There are several reasons. The first is perhaps a quixotic frustration centering on its meaning. Its etymology stems from the Greek words *a* (ἀ) and *theos* (θεός). Separately, the latter means *god* and the former is a negatory. In combination, they mean godless. Yet this pure denotation belies a heterogeneity of connotations. Consider surveys that attempt to measure personal declarations of faith. Is there a distinction between atheism and non-religious beliefs? Can one affirm atheism yet hold spiritual beliefs? Does marking "no religion" on a form equate to atheism? Do surveys rightly distinguish atheism from agnosticism? The boundaries of atheism are decidedly ambiguous, as there is no common, widely-held, cross-cultural definition of the term.[41]

SIMPLICIO: I understand the dilemma. Many faiths likewise encounter confusion among the teachings of their various denominations, schools, and sects.

SALVIATI: Consider also notional interpretations. The devout might paint a dark, valueless, amoral, all but the most messianic portrait of atheism. Its practitioners, however, describe an enlightened, meaningful, value-laden, personally rewarding philosophy, one unencumbered by mythical fabrications. Antagonists might counter by equating atheism with secularism, contending the latter propagates the former. Thus the term *atheism*, rather than acting as a neutral marker, inevitably serves to provoke a rash of value judgments.

SIMPLICIO: To avoid the knottiness of meaning, perhaps we can adopt an intuitive definition not uncommon in contemporary monotheistic societies: disbelief in the prevailing deity. That is, rejection of the one true God who created the world; the God who periodically intercedes to influence worldly events; the God who knows you exist, cares about you, and responds to your prayers.

SALVIATI: Yes, that is a starting point. Yet a theistic deity is but one thing atheists do not believe in. They likewise commonly gainsay psychics, mystics, spirits, spiritualists, astrologers, sorcerers, fortune tellers, demons, trolls, gnomes, ghosts, goblins, elves, dwarves, pixies, sprites, nymphs, fairies, gorgons, harpies, ogres, dragons, leprechauns, minotaurs, angels, valkyries, specters, phantoms, poltergeists, shamans, seers, unicorns, cyclopes, werewolves, vampires, zombies, Yetis, and Loch Ness monsters.

SIMPLICIO: That is a lengthy list.

SALVIATI: It is longer still. Atheists also routinely abjure magic spells, pyramid power, oracles, healing crystals, energy vortexes, omens, incantations, enchantments, conjuring, clairvoyance, augury, sorcery, curses, hexes, runes, talismans, numerology, horoscopes, the occult, water-witching, tea leaves, witchcraft, wizardry, evil eyes, channeling, blessings, prayers, rain dances, apparitions, rogations, New Age therapies, reiki healing, tarot cards, reflexology, homeopathy, telekinesis, and paranormal sightings. It ought to be clear from this litany that atheism is too restrictive a word.

SIMPLICIO: Your list suggests it is too restrictive by far.

SALVIATI: I neglected to mention that many atheists also distrust post-modernist philosophy, due to its allergy to objective truth.

SIMPLICIO: My faith, too, distrusts post-modernism. Rather than atheism, however, some suggest the term *agnosticism*, the belief that the existence or non-existence of deities is unknowable.

SALVIATI: Yes, but I find this term equally unsuitable. Agnosticism is an indecisive stance. It ignores the reality that the faith-based thinking method of religion, having failed for millennia to supply compelling evidence for its claims, is unlikely ever to do so. Further, it discounts the ability of evidence-based thinking to continue the expansion of its explanatory power, one day fully illuminating all the salient components of our world.

SIMPLICIO: So you find both terms, atheism and agnosticism, unacceptable?

SALVIATI: Yes, I do. And both suffer an added defect, arguably their most severe. They describe what a person does not believe, but say nothing of what they do believe. If any term defines a person, it should reflect what he or she believes. Yet here resides a void. Those who accept critical and evidence-based thinking as the basis for comprehending the world lack a cohesive term to describe themselves. Many have endeavored to coin an apt label. Examples, to name but a few, include naturalist, rationalist, freethinker, secularist, and skeptic. Perhaps I set too lofty a bar, but for a variety of reasons, I find all these words inadequate to the task of representing the philosophy with importance, beauty, and resonance. A term is required that portrays critical thought as an island of objectivity amid a dark and murky sea of superstition, as a lighthouse whose beacon both illuminates and dispels a myth-filled world. I must, I'm afraid, leave it to more erudite and creative minds to find an apt term.

SIMPLICIO: You do set a high bar. The wait might be long.

SALVIATI: Yes, that likely is the case. Until that time, however, I will remain with *humanist*: one who bases their factual and ethical decisions on reason and science; one who advocates the liberal human values of freedom, justice, and tolerance; and one who rejects the existence of supernatural authority.

SIMPLICIO: You might wish to remain with the label *atheist*. Many will not comprehend the traits or subtleties of humanism.

SALVIATI: You probably are correct. Your words remind me of an astute insight on the word atheist: "When you understand why you dismiss all the other possible gods, you will understand why I dismiss your god. I contend we are both atheists. I just believe in one fewer god than you do."[42]

SIMPLICIO: Clever. The quote conveys a certain truth. But consider a higher truth, "I am the Lord thy God."[43]

SALVIATI: Again, touché.

SIMPLICIO: I enjoyed our conversation today. Blessings, Salviati.

SALVIATI: Until tomorrow, Simplicio.

SIMPLICIO: (to himself, exiting the café) Indeed, all are atheists, apart from those who embrace the one true God.

SALVIATI: (to himself) I wonder if Simplicio recognizes that he, too, is an atheist, in the eyes of many non-Christian religions.

*End of the Fifth Day*

{ Day Six }

# On Abuse Of Labels

*We have to repudiate this idea that beliefs can be sanctified by something other than evidence and argument.*[44]

Sam Harris (1967-)

American Neuroscientist

SIMPLICIO: Hello, Salviati. I have been impatiently awaiting your arrival. When I awoke this morning, I found myself contemplating our discussion on labels. I thought I might mention a label I particularly dislike, one that has entered our conversations on multiple occasions.

SALVIATI: Please disclose it, Simplicio. It can be our topic today.

SIMPLICIO: It is the word *theory*. You have enlightened me on its scientific meaning as a likely explanation of phenomena, founded on extensive and compelling evidence derived from experiment. And you described each theory as a provisional truth.

SALVIATI: I might say a provisional *understanding* rather than *truth*, but yes, you capture the essence of the definition. How does the term *theory* disturb you?

SIMPLICIO: I do not challenge its definition in the context of, what you like to call, evidence-based thinking. But I admit I sometimes flinch when the term is applied in a religious context.

SALVIATI: In what way?

SIMPLICIO: Some individuals, skeptics primarily, refer to God's creation of the world as a *theory*. Its use suggests the narrative in Genesis is a tentative truth, one that one day might be unseated by a newer truth. Yet this cannot be. The Lord, in no uncertain terms, tells us He created both the heaven and the earth.[45] He further elucidates how he brought the world and all its beings, day by day, into creation.[46] As Holy Scripture expresses the word of God, its truths cannot be transient; they are eternal.

SALVIATI: I understand your dilemma. It might come as a surprise that I agree with you.

SIMPLICIO: Indeed, it does. You acknowledge Holy Scripture as the true and inerrant word of God?

SALVIATI: No, I merely agree the word *theory* is misapplied when used with religion. Theory, as a scientific term, carries a robust meaning, one we already have discussed. More importantly, it derives from a mode of thinking that differs decidedly from religion. A theory forms from a distillation of both confirmed and rejected hypotheses, each formulated on testable predictions. It is the culmination of a series of steps, each designed to yield credible, defendable conclusions. It results from a process both transparent and robust. As religion lacks this process, the term *theory* never should be applied to doubtful knowledge garnered from deistic authority.

SIMPLICIO: It seems we both agree the word *theory* should not be applied to religion, but from different rationales.

SALVIATI: You are correct. And you might advise your religiously-inclined brethren to note this fact: It is they who most often misuse the word. Creationists, in particular, creatively employ the word *theory*. By leveraging scientific anomalies and culling unwelcome data, creationists assemble a hodgepodge of miscellany, apply a glossy sheen of specious plausibility to the contrivance, then label it a *theory*. Their objective, it should go without saying, is twofold. First, to elevate their fatuous claims to a higher bar, to the objective standard of science. And second, to feed the dogged enthusiasm of the disciple who, head afire from zeal, apprehends an obligation to impose his beliefs on the less enlightened. Perhaps you could educate your creationist apologists on both their misuse and abuse of the term.

SIMPLICIO: Creationists inspirit legions of the faithful, but I will acknowledge a private truth to you. Their very actions suggest that, contrary to God's teaching, faith alone is insufficient for salvation. They require temporal signs to justify their belief.

SALVIATI: I have never heard it so expressed. You have enlightened me. Allow me, in turn, to suggest a further misuse of *theory*.

SIMPLICIO: Please, do.

SALVIATI: When circumstance suits, religious proponents practice a charade opposite to the creationist dodge. When confronted with an unwelcome theory from science, they label it "just a theory." The context and tone of the utterance serve to denigrate the theory as something sketchy and incomplete, something capricious and whimsical, something scarcely worth one's attention. The intent, of course, is to invite ridicule and dismissal of distasteful theories, those that belie sacred doctrine.

SIMPLICIO: Yes, I have heard such things said.

SALVIATI: Religion's bipolar use of the term *theory* to both exalt its own beliefs and belittle those of science is not uncommon. It employs other vocabulary in a like-minded vein. Consider the words *belief* and *faith*. Religion routinely applies the terms to glorify its dogma, yet applies the same words to cast doubt on science. The theory of evolution by natural selection, for example, is branded "just another belief." Atheism is labeled "just another faith," albeit a false one. Religion's opportunistic use of the terms is understandable but erroneous. A theory produced by science is not just another belief. Atheism is not just another faith.

SIMPLICIO: I believe I know why, but please continue.

SALVIATI: The answer, once again, stems from the underlying thought processes of science and religion. *Belief* and *faith* arise from the faith-based thinking of religion, a process based on revelation, scripture, and the whims of an invisible entity. But a scientific *theory*, as we have discussed, is not built on ingrained, unquestioned faith. It is not the product of deference to Holy Writ, prophetic assertions, or divine authority. It is not established, sustained, or vindicated "because a book says so." And it is never worshipped. It issues from the rigorous, transparent, and evidence-based process of science. As such, *theory*, *hypothesis*, *science*, and analogous terms arise from a paradigm that differs decidedly from religion - a paradigm for comprehending the cosmos and its contents from detailed observations and testable predictions. Such predictions, in turn, provide an unmatched ability for discerning truth from fallacy.

SIMPLICIO: And when truth is proved to be fallacy?

SALVIATI: In such instances, I contend, practitioners of evidence-based thinking act rationally; they discard the refuted theory. But the case differs for faith-based thinkers. Another bipolarity emerges. Consider the distinction between disproof of an unpalatable *scientific* claim versus disproof of a *religious* claim. When the former is disproved, the religious mind interprets disproof as confirmation that "it was just another theory, one that we knew never could be true." Yet when science disproves a religious claim, the same pietistic mind never disavows its stance. In fact, refutation all too often provokes reinforcement of the superstition. The inability to disavow, a trait foreign to science, is a hallmark of *belief* and *faith*.

SIMPLICIO: Your point is taken. But when the truths radiate from the Lord, there is not the slightest need to disavow.

SALVIATI: Still, I find it curious that religion employs *belief* and *faith*, terms so integral to its own dogma, to deprecate the findings of science. It is as though religion itself understands the terms convey little substance.

SIMPLICIO: To the faithful, the terms carry immense substance. They transmit an essentiality that, if I may say so, eludes you.

SALVIATI: Perhaps. But the saga is similar with *evidence*, *proof*, *truth*, and related vocabulary. Religions often employ the terms in attractive catchphrases: "six proofs of God's existence," "evidence for God's hand in creation," and "the truth of the Savior's love." Sadly, the shibboleths often erect an effective smokescreen to obscure, mislead, and bemuse. Yet science is able to penetrate the haze. Closer scrutiny invariably reveals oddities - vague predictions, absence of detail, lack of transparency, contrived evidence, selective use of data, imaginative interpretations, dissembling. The oddities come as no surprise, as all three terms - *evidence*, *proof*, *truth* - to the religious mind, share a common definition: an agreed interpretation of an often cryptic passage from a centuries-old text.

SIMPLICIO: I cannot concur with your conclusion. Yes, the Bible is centuries-old, but its truths remain invariable. The faithful possess the key to decipher the cryptic, a topic worthy of discussion in its own right. But remaining with today's theme, there must exist other words whose definition or use differs between religion and science?

SALVIATI: Yes, many. Some of greater purport, some of less. Three instantly come mind. *Skeptic*, *natural*, and *significant*.

Religion often labels a skeptic as *one who denies*. Science, in contrast, refers to a skeptic as one who refrains from forming an opinion on sketchy evidence. Skeptics eschew comforting, convenient, diplomatic, and impulsive judgments. Equally important, they value the method by which a conclusion is reached more highly than any particular outcome. Science values skepticism as a key component of critical thought.

SIMPLICIO: It is a fair distinction. What issue do you see with *natural*?

SALVIATI: On its own, none. The issue arises in contrast to *supernatural*. Theists all too often deploy the terms to segregate the realm of religion from science. They attempt, of course, to isolate the *supernatural* as the exclusive domain of religion, free from the scrutiny of scientific inquiry. The implication is that science should constrain itself to the *natural* world, where its mindset and methodology pose less threat to religious claims. But this is a false distinction. To science, the natural world is the whole world. If there were supernatural phenomena, they, too, would be part of the natural world - to be equally discovered, examined, and verified (or disproved) as any other phenomena.

SIMPLICIO: It is an interesting distinction, but one that theists unlikely will accept. As we have discussed, the faithful often find scientific inquiry unwelcome.

SALVIATI: I believe you are correct. The attempt to insulate religion from science is an ancient one, unlikely soon to be resolved. Allow me, however, to proceed to the third term I mentioned.

SIMPLICIO: Yes, please. What is the significance of *significant*?

SALVIATI: In religion, *significant* denotes *important*. In science, it simply refers to an event unlikely to occur from random chance. It conveys no connotation on whether the event itself is noteworthy.

SIMPLICIO: The religious connotation of *significant* coincides with the general understanding of the word. Very few, I suggest, would be aware of its scientific overtone.

SALVIATI: That probably is the case. Much opportunity remains for science to make itself better understood. Communication, an art in which you and your religious colleagues excel, is not a forte of science.

SIMPLICIO: Thank you for the compliment, Salviati. I see the hour, once again, is late. I am happy you agreed with my original premise, that the word *theory* is misused when applied to religion. I admit you expanded my understanding of the breadth of its misuse, and of the other terminology you mentioned.

SALVIATI: Thank you, Simplicio. It is kind of you to say so. Shall we adjourn until our usual time tomorrow?

SIMPLICIO: Yes, certainly. I have another challenge I wish to put to you. Blessings, Salviati.

SALVIATI: I look forward to it. Goodbye, Simplicio.

SIMPLICIO: (to himself) Yes, creationists should refrain from using the word *theory*; the guise undercuts their arguments. They should rely on faith alone.

SALVIATI: (to himself) In the conversation today, did I sense a heightened level of introspection from Simplicio?

*End of the Sixth Day*

Diagram 4.1 A selection of vocabulary as used by proponents of religion and science.

| Term | Religion | Science |
|---|---|---|
| Belief | What God tells me. | A codeword for not wanting to know what is true. Antonym: science. |
| Faith | What God tells me. | Beliefs designed to thwart scrutiny and testability. Like mythology, but less entertaining. |
| Natural | The physical world. | The entire world. |
| Supernatural | The metaphysical world. Synonym: God. | See *natural.* |
| Theory | An unwelcome scientific claim that challenges the truths of faith. | A provisional truth supported by robust evidence. |

{ Day Seven }

# On Negation

*Prove all things.*[47]
The Bible

SIMPLICIO: Hello, Salviati. Are you arriving late today, or am I simply overanxious to put my next challenge to you?

SALVIATI: My apologies if the former, but I hope it is the latter.

SIMPLICIO: I trust my challenge will not disconcert you.

SALVIATI: I am all attentive. Fire away. Your inquiries help to hone my thoughts, as mine do yours, I hope.

SIMPLICIO: No doubt you encounter, from time to time, someone from the faith who wishes to testify to you?

SALVIATI: Occasionally, yes, I am the target of such encomiums.

SIMPLICIO: Therein lies my challenge. I am curious how you react to the true believer. What do you say to someone who has met the risen Lord Jesus Christ; who has received the anointing of the Holy Spirit; who has walked with God at his side; and who, through the grace of the Lord, has withstood the temptations of the antichrist? What do you say to the disciple who extols their faith in the one true God, He that made man in his His image, and woman from man?

SALVIATI: Your challenge is well-articulated. When the question is sincere, as in your case, I offer a threefold response. First, I say belief is a function of time and place. If you had been born in Sudan, you would be asserting the same truths about Muhammad or Allah that you do about Jesus. If you had been reared in Punjab, you would be avowing, just as strenuously, the dogma of Ik Onkar. If you had been raised in ancient Rome, you would have reserved your highest reverence for the sky god Jupiter. If your family had lived among the pre-Columbian Maya, you would have shouted hosannas to Kinich Ahau. If your parents had resided in pre-Christian Scandinavia, you would have acclaimed the divine wisdom of Odin.[48]

SIMPLICIO: You are suggesting belief is relative?

SALVIATI: Yes, I am. Faith derives, in large part, from the prevailing beliefs of a particular geography and era - the where and when of your birth. Throw in a few other variables - the beliefs of one's parents, one's socioeconomic class, even one's gender - and one's faith is all but determined. Alter these variables at nascency, and you alter one's faith. Few, in fact, choose their faith; it is thrust upon them. Note the upshot. Faith derives from no deistic authority, from no set of scriptural truths. It simply is a consequence of birth. This explanation, by the way, clarifies many aspects of religion, such as its tendency to cluster geographically.

SIMPLICIO: What do you mean?

SALVIATI: In the sparsely-populated, geographically-separated settlements from millennia ago, each society spawned its own deities, its own beliefs. Both deities and beliefs tended to remain near their place of invention, affording spiritual sustenance to their inventors. Travel one-hundred kilometers, and you would have encountered a new community, a new deity, and a new dogma. Travel another stretch, and you would have happened upon yet another society, yet another divine being. This pattern, each religion adhering to its place of origin, was the predominant model for much of the history of our species.

SIMPLICIO: That seems reasonable. In ancient times no facility existed to disseminate information across diffuse geographies.

SALVIATI: That is correct. Now let's extend the paradigm. Consider the content of holy texts. The Bible is an exemplar. The text makes no mention of coeval societies outside its limited purview. It recounts the acts of no peoples living in lands more than a few hundred kilometers distant. Why? Because its authors merely chronicled the local lore of their own restricted geography. All sacred texts attest to this fact. They transmit no celestial wisdom avouched to a broad, ecumenical populace. Each narrative simply reflects the myopic beliefs of a parochial locale.

SIMPLICIO: But, as you affirmed, no mechanism permitted ancient peoples to communicate a shared faith over sizable distances. Thus, no author was able to articulate a worldwide perspective. It was only later, as populations expanded, that religions spread beyond their origins, carried by migration, trade, and conquest.

SALVIATI: Exactly. The then-extant authors possessed no such means. But someone did.

SIMPLICIO: Who?

SALVIATI: The deities. The divine beings invented by each provincial society. The ephemeral entities who, as avouched by their inventors, possessed infinite knowledge and limitless talents. They commanded every conceivable means. No obstacles thwarted their will. They could have imparted catholic truths to all their creations, from pole to pole, concurrently. Yet none did. None spoke to all peoples worldwide. None revealed a consistent creation myth. None proffered an identical ethical code. None imparted the same assemblage of universal truths. None was everywhere affirmed as the one true god. Not one. Each of the countless chimerical deities pandered only to the inhabitants of a narrow locality, to the same denizens who first had forged the deities. To reaffirm, each holy book simply documents the superstitions of its finite locale.

SIMPLICIO: I will not say I agree with you, but please proceed.

SALVIATI: I will move on to my second point. The human mind is highly susceptible to what it wants to believe, a susceptibility driven from various psychological wants, among them a search for meaning, a need for emotional connection, and a craving for security. When theologies arise that sooth our wants, we all too often embrace their pronouncements with little scrutiny. We then label our unquestioned beliefs as truths, ardently defending them to protect our improvised mental world. From this tendency, it is simply too easy to believe what we want to believe, what we claim to be true, rather than what can be shown to be true. The impulse to believe is just too great.

SIMPLICIO: You suggest the human mind creates its own gods?

SALVIATI: I do. And from this premise, I contend the conclusions obtained from faith-based thinking - from entreaties to and revelations from an unseen deity - simply cannot be trusted. They possess no credible ability to explain the world in which we live. This accentuates one of the unique aspects of evidence-based thinking. It helps us find what really is true, not what we merely wish to be true.

SIMPLICIO: You seem to devalue the human mind.

SALVIATI: Not at all. I consider it the current culmination of evolution, one capable of remarkable feats. Yet it is not perfect. Disparate disciplines attest to its susceptibility, including its inclination toward supernatural beliefs. Beyond the wants just mentioned - for meaning, security, and emotional connection - psychology offers various other hypotheses. One is the need to suppress cognitive dissonance, the psychological discomfort individuals experience when they are forced to confront ideas or events that contradict their own beliefs, opinions, or values. All too often, an overwhelming yearning to repress such dissonance engenders an insular mindset, one that dismisses any fact that contradicts its worldview.

SIMPLICIO: Yes, I am familiar with cognitive dissonance. Apart from psychology, do other fields support your claim regarding the susceptibility of the human mind?

SALVIATI: Yes, many do. Paleontology, sociology, neuroscience, and evolutionary biology - among other sciences - tender both competing and complementary conjectures. Some concern the survival instincts of our paleo-ancestors and the structure of their societies. Others derive from the physical and chemical make-up of our brains. All attest, in one form or another, to a tendency of our species to establish preferred beliefs. And once those beliefs are installed, to sustain them even in the face of contradictory evidence.

SIMPLICIO: It is clear you have thought these matters through. What is the third point you would make?

SALVIATI: I would ask the apostle to dispassionately assess religion's track record in explaining the world. I would request he evaluate its myriad predictions of the second coming of Christ, of the end of the world, and of other religiously significant events. What percentage of the predictions proved to be true? If his analysis is unbiased, he will arrive at a single, salient conclusion: Religion's history of successful predictions is abysmal.

SIMPLICIO: It is true Christ has not yet returned in physical form, but you cite few other examples to support your contention.

SALVIATI: Then let me cite more of the specious pronouncements from creationism, intelligent design, and other religiously-inspired pseudosciences. The thinness of moon dust attests to a short period of accumulation, disproving claims of a moon billions of years old. Human fossils found in ancient geological strata - strata older than the origin of the first humans as claimed by science - refute the theory of evolution. The second law of thermodynamics, which suggests disorder increases over time, cannot account for a structured universe. No present-day experiments have demonstrated that species evolve. The absence of fur on Inuit peoples disproves evolution. Earth stood vertically on its axis before the Biblical flood, thereby explaining the "truth" of some obscure Biblical passage. Human footprints have been found side-by-side with dinosaur imprints. The Bible documented continental drift, "for in his days was the earth divided,"[49] long before its discovery by science. God created light "in transit," thus all measurements to distant stars calculated from the speed of light and its redshift are incorrect.

SIMPLICIO: I was not aware there were so many.

SALVIATI: Shall I reel off still more? A smaller than expected number of neutrinos emitted by nuclear fusion within the sun disproves an oldish star. The relatively high level of helium within the earth's crust affirms a newish planet. Conversion therapies cure homosexuality. The lifespan of Neanderthals was 300 years. Vapor in the firmament was the source of the water for the Biblical deluge.[50] Ice crystals in equatorial rings around the earth were the wellspring of the rains that led to the same torrent. The occasional protein shaped like a cross is proof of Jesus. There are no transitional evolutionary forms. Isotropic properties of the universe attest to the earth's placement at the centre of the cosmos. Darwin, on his deathbed, recanted his theory of natural selection. Catastrophic plate tectonics caused continental drift to occur in months, not over millions of years, shaping the planet we see today. All life forms require a membrane, from bacterial cell walls to human skin, to enclose their body parts; because life cannot form without a membrane, and a membrane cannot form without life, the existence of membranes proves all living creatures originated from God.

SIMPLICIO: Enough, please. You have established your point.

SALVIATI: Very well, but note a common theme among the assertions. Their absurdity is surpassed only by their sophistry. Purblind proponents construct their deceits from endless guile: by altering or cleansing distasteful evidence; by discrediting or dismissing unwelcome facts; by rejecting or ridiculing disproofs of entrenched tenets; by constructing assertions to be untestable, hence irrefutable; by elevating belief above empirical evidence; by hoisting piety over cogent argument; by ascribing coveted myths to theistic authority; by partitioning faith into a separate realm, one free from scientific attack; or by offering a trite *Goddidit* as explanation for any baffling phenomena.

SIMPLICIO: It is clear you hold the proponents in low esteem.

SALVIATI: Yes, I do. Note, too, the charade extends further, to interminable reinterpretation. Consider this scenario. The zealot declares the earth to be five thousand years old, in literal agreement with Holy Writ. As science advances, geologists suggest an older age, as they did in the nineteenth century, of 200 million years. The zealot, unable to produce a thread of evidence for his contention, now finds himself in a bind. He is reluctant to disavow his myth, but can locate no scriptural basis for an older earth. So he lands on a stratagem. He abandons his long-held absolutist belief, in this instance, that the six days of creation reflect six twenty-four-hour periods. He ditches his fable, supplanting it with a newer fiction, ascribing to each Biblical day an interval of roughly thirty-five million years. Thus, six days of creation times 35 million years per "day" equates roughly to 200 million years. Presto. His artifice wondrously restores Biblical and scientific alignment.

SIMPLICIO: You must admit a certain cleverness.

SALVIATI: Or guile. Either way, the makeshift alignment is temporary. It soon evaporates.

SIMPLICIO: How so?

SALVIATI: Simply put, science progresses. In the mid-twentieth century, science revised its estimate for the earth's age. From advances in nuclear physics, science conjectured an age of 2 billion years. The zealot, in response, recast his fraud. He quietly and unashamedly discarded his metric of thirty-five million years per Biblical day, touting forthwith a newly-minted truth, a span of some 330 million years for each doctrinal day. In the twenty-first century, the cycle repeated, with some contending that each scriptural day now equates to 750 million years, forcing Biblical alignment to science's current estimate for the age of the earth: 4.5 billion years.

SIMPLICIO: You fault the zealot for reinterpreting the age of the planet, yet not the scientist who revises his estimate even more frequently?

SALVIATI: Correct. The scientist continuously seeks to expand and refine his understanding of the cosmos, pursuing evidence wherever it might lead. The zealot simply wishes to shield a deep-rooted belief, in this instance, that the earth was created over six magical days. His ploy - to continually refashion the interval of each enchanted day, declaring every incarnation a hard-and-fast divine truth - rebuts his own myth.

SIMPLICIO: You imply the zealot condemns himself to perennial reinterpretation as science advances.

SALVIATI: Often, that is the case. But the zealot learns from his machinations, honing his stratagems. Today, he often abandons the arithmetic façade entirely. He declares that a day in

Genesis reflects no precise interval of time but merely is allegorical, revealing a true but timeless act of the Lord. The act cleverly but ignominiously disavows a core, literal belief of his faith, one held for centuries if not millennia, replacing it by an endlessly malleable interpretation that, from his perspective, never can be disproved. How convenient.

SIMPLICIO: Your words suggest there is no line in the sand beyond which the faithful will not cross to shield their beliefs.

SALVIATI: Precisely. Their sham predictions, their spurious reinterpretations - all arise from fanciful thinking, all contrive to uphold the tenets of a single faith, their own. Yet all have been either unceremoniously disproved or rebuked from want of evidence. Religion would be better off asserting its claims from the flip of a coin. Its rate of successful predictions would be both higher and more honestly obtained.

SIMPLICIO: You will know I cannot accept that answer.

SALVIATI: Yes, I know. But I ask you to compare the results offered from evidence-based thinking. It is not perfect. It produces errors. It ofttimes is lethargic. But its reliance on testable predictions provides a nonpareil capability to extricate fact from fabrication. In every instance where religion has described a testable aspect of our world, and evidence-based thinking has put forth a rival explanation, the latter has proved to be more correct.[51] In every case. I repeat, the track record of religion, over the millennia of its existence, is dreadful.

SIMPLICIO: But it is true, is it not, that of the explanations and predictions put forward from science, a fair number turn out to be incorrect?

SALVIATI: Yes, many do. But as we have discussed, scientific claims, at most, represent provisional truths. Each might be revised or discarded as new evidence comes to light. This ethos stands in stark contrast to religion, which portrays its declarations as eternal and immutable. Scientific refutations of spiritual truths thus become, to the religious mind, problematic; they expose the charade of faith-based assertions.

SIMPLICIO: Perhaps the erroneous religious claims, explanations, and predictions emanate from false prophets? Perhaps it is only from listening to the one true God that we divine irrefutable truth?

SALVIATI: Perhaps, but such a solution hints at cozenage. It is much too easy to relegate a disproved religious prediction to a false prophet. It permits the zealot to readily assert that, "if the claim was false, it must be that we we were mislead by an emissary of Satan." Ascribing unwelcome facts to a test of faith is equally duplicitous. The intent of such chicanery is obvious to the skeptic. It serves merely to dismiss any fact or opinion that contradicts one's enshrined beliefs.

SIMPLICIO: You deny the existence of false prophets?

SALVIATI: Only to the same extent of denying real prophets. We simply have to dispel the notion that the words *I believe* form a get-out-of-jail-free card, one that liberates its holder from any inconvenient or unwanted truths. We must oppugn *faith* when it serves as an enchanted token to shield its owner from any evidence that disproves a cherished belief, when it acts to expunge findings that discomfort the intransigent believer, or when it marshals its forces to decry objective inquiry.

SIMPLICIO: And replace the get-out-of-jail-free card by what?

SALVIATI: First and foremost, by mental freedom. By jettisoning the weighty, constricting baggage of myth and mysticism. By eliminating the need to argue away the contradictory assertions and illogical tenets of theistic beliefs. By expelling the cognitive dissonance that arises when evidence assails dogmatic convictions, yielding a psychological benefit not afforded the theist. By relishing the self-reliance that emerges from taking greater control of one's destiny, rather than subjugating oneself to the whimsical acts of an invisible being. By embracing new paradigms of thought and inquiry, well beyond the artificial restrictions imposed by theistic authority. By reveling in the clarity and elegance of a fact-based world. By founding one's decisions on critical thought and empirical evidence which, in turn, drive credible, defendable actions.

SIMPLICIO: You cite a lengthy list.

SALVIATI: Yes, and there is more. By inspiriting a morality well above the meager bar of Christianity. By advocating social justice beyond a faith's arbitrary assignation of one's lot in life. By magnifying the joy in family, friendships, and life itself, driven from the comprehension our time on Earth is transitory. By sustaining the secular traditions that interlace a community. By role-modeling the nobler virtues of our species. By contemplating art, poetry, and prose, inspiring us to look beyond ourselves. And by standing in awe of both the beauty of our world and the sublime majesty of the cosmos. Shall I cite still more?

SIMPLICIO: No, thank you, your list is sufficient. But if I may say so, some elements feel a bit academic.

SALVIATI: Then let me be more practical. For millennia, humanity beseeched supernatural entities for bountiful harvests, robust health, protection from the elements, and much more. But who delivered the goods? Who electrified our cities, reduced pain in childbirth, invented childhood vaccines, preserved food through refrigeration, solved navigation at sea, devised buildings to withstand earthquakes, engineered drought-resistant crops, churned out antibiotics, crafted structures to prevent floods, created lenses to improve vision, subdued plagues, lengthened our lifespans, transmitted voices around the globe, comprehended human psychology, expanded crop yields, curtailed childhood mortality, and constructed motors to ease physical labors? Who produced these results, among countless other practicable contributions? It was science. It was science, and still is, that best satisfies human wants and needs. It is science that produces a constant stream of discoveries, driving myriad tangible benefits to humankind. Why? Because science works. It works for all of us. When you stream sermons to your congregation, it works for you too. But it was science that delivered the technology, not your deity.

SIMPLICIO: You contend, then, from your enumeration of academic and practical contributions, that science enables happier lives?

SALVIATI: I do, for an increasing percentage of the planet's inhabitants. It is science, coupled with liberal democracy and the rule of law, that drives the betterment of our societies. All free from magical thinking.

SIMPLICIO: Noble words, Salviati. I will ponder your outlook. Now, I notice the sun is approaching the horizon. Shall we adjourn to the terrace and watch its descent?

SALVIATI: By all means.

SALVIATI: (to himself) Might Simplicio reexamine his mystical truths? I wonder.

SIMPLICIO: (to himself) Salviati, he loves to taunt. Yet his speech is persuasive. I wonder.

*End of the Seventh and Final Day*

*And on the seventh day, Salviati and Simplicio suspended their discourse, and they rested.*[52]

Salviati's threefold response. **Diagram 7.1**

| What a Religionist Says | What a Skeptic Discerns |
|---|---|
| ① There is one true faith. | Faith is a function of birth. The geography and era into which one is born determine religious affiliation. The prevailing deity plays no role. |
| ② I embrace the one true faith. | The mind is susceptible to what it wants to believe. Science mitigates this tendency through testable, evidence-based assertions. |
| ③ My faith speaks truth. | The ability of religion to explain the world is deplorable, its predictions of literal and immutable truth regularly disproved. |

# Epilogue

*Man appears to be the missing link between anthropoid apes and human beings.*[53]
Konrad Lorenz (1903-1989)
Austrian Zoologist

Salviati and Simplicio first met in 1632, but the seed of their friendship began twenty-two years earlier, in 1610. In that year, Galileo Galilei (1564-1642), the Italian astronomer and physicist, published *The Sidereal Messenger*, a work in which he described his telescopic observations of the moon, and of the planets Venus and Jupiter. Among his discoveries, he detected four moons in orbit around Jupiter. To twenty-first century eyes, Galileo's explorations advanced our understanding of the universe. To the contemporaneous mindset, however, his findings provoked a hostile reaction, one orchestrated from the decidedly unscientific ethos of the Catholic Church.

In the seventeenth century, the Church still subscribed to geocentrism, a model put forward 1500 years earlier by the Greco-Egyptian natural philosopher Ptolemy (c.90-c.168). The model placed a non-moving earth at the center of the universe, with the sun, moon, stars, and other planets orbiting the earth.[54] This heavenly arrangement, as construed by the Church, affirmed

God's will - the placement of humanity at the locus of divine creation, our species serving as the cynosure of the universe. Scriptural interpretation sustained the orthodoxy: "the world also shall be stable, that it not be moved"[55]; "the world also is stablished [*archaic*] that it cannot be moved"[56]; and "the world also shall be established that it shall not be moved".[57]

To the dismay of the Church, Galilean moons in orbit around Jupiter challenged the geocentric model. Their presence indicated that not all celestial objects orbit the earth. Worse still, satellites circling a distant planet signaled the correctness of heliocentrism, a competing model of the cosmos put forward in 1543 by the Polish mathematician and astronomer, Nicholas Copernicus (1473-1543). The heliocentric model places the sun, not the earth, at the center, with a moving earth orbiting its home star.

Unwilling to question its absolutist belief, the Church took action. The Inquisition, an institution charged with the eradication of heresies, declared heliocentrism heretical as it explicitly contradicted Holy Scripture. Galileo was ordered to refrain from teaching or defending heliocentrism, at threat of stronger action from the hands of the Inquisition, including the very real risk of torture.

Admonished but by no means silenced, Galileo cautiously, gradually, and scientifically constructed his apologia. In 1632, he published his seminal work, *Dialogue Concerning the Two Chief World Systems*. His opus openly argued for a heliocentric model of the cosmos. He presented his arguments as a sequence of conversations over four days, each discourse debated among three antagonists: Salviati, a lawyer, who champions the Copernican system; Simplicio, a philosopher, who advocates the Ptolemaic system; and Sagredo, a neutral observer. It did not help that the name Simplicio, who defended the position of the Church, purportedly evoked the Italian word *semplice*, meaning *naive* or *simple-minded*.

Again, the Church acted, this time with greater force. It

placed *Two Chief World Systems* on its *Index of Prohibited Books*, a list of publications banned by the Church. It arrested, tried, and convicted Galileo of heresy. It sentenced him to imprisonment, but later reduced the punishment to house arrest, a penalty that remained in force to the end of his life. The Inquisition further instructed Galileo to read the seven Penitential Psalms[58] as an expression of sorrow for his sin. Perhaps they thought the penitent act would preserve their errant portrait of the universe. It did not. In the end, the Church failed to defend its absolutist, dogmatic belief in the Ptolemaic model. Invocations of deistic and temporal authority afforded it no refuge. Far from it. Mounting evidence in favor of heliocentrism ultimately disproved both geocentrism and Church doctrine. In 1835, the Church removed *Two Chief World Systems* from its *Index*. In 1992, it formally cleared Galileo of any wrongdoing, 360 years after the publication of his most celebrated work.

Salviati and Simplicio, in reprising their roles as the antagonists of this tome, pay homage to the author of their original debates from four centuries ago.

---

Miscellany

Catholicism was far from the only religion to condemn heliocentrism as a threat to its faith. The German theologian Martin Luther (1483-1546), as case in point, trumpeted the same sentiments from his Protestant pulpit, reportedly declaring Copernicus a fool for challenging Holy Writ.[59] The emphasis in this text on the Catholic Church arises from its interactions with and treatment of Galileo.

# On Absolutes Anew

*The existence of trousers proves that God meant us to be bipeds.*[60]
David Austin (1935-2005)
British Cartoonist

SIMPLICIO: Hello, Sagredo. It's been some time since we've seen you.

SAGREDO: Hello, Simplicio. Hello, Salviati. Yes, I just returned from a trip. It's nice to see both of you again. Did I miss anything while I was away?

SIMPLICIO: Salviati has been entertaining me with his dialogues. You know the type of discourse in which he likes to engage.

SAGREDO: Yes, I know the type, argumentation that frequently induces eyes to narrow and lips to purse. Salviati, have you been disrupting Simplicio's cognitive consonance?

SALVIATI: I plead guilty to the charge.

SAGREDO: I am not surprised. But that reminds me, Salviati. You should have been with me the other day. I encountered a situation that would have piqued you. I overheard a conversation between two individuals. The first, citing various forms of evidence, claimed the earth is 4.5 billion years old. The second asserted an age of 5,000 years, in alignment with Biblical scripture. The first continued to press his case but, in the end, the second posed a single question. Can you demonstrate with absolute certainty the earth is 4.5 billion years old? The first could not, so the second declared victory in the debate.

SALVIATI: Yes, I wish I had been there. The question you overheard is deceptive. Permit me to explain. I would commence by pointing out ...

SIMPLICIO: Salviati, excuse me for interrupting, but please allow me to respond.

SALVIATI: By all means.

SIMPLICIO: As Salviati was about to say, the question posed by the second individual suffers from multiple fallacies. First, the interrogator misunderstands the evidence-based method of science. Science forms its theories from experiments, each designed to confirm or refute specific hypotheses. A theory, at its simplest, is a set of confirmed hypotheses forming a likely explanation of the phenomenon under study. Behind the theory lies a preponderance of evidence which substantiates the explanation. Yet a theory always is provisional; it might be revised or overturned as new discoveries come to light. Thus, a demand for "absolute certainty" misconstrues what science is able to offer.

SAGREDO: Simplicio, you amaze me.

SIMPLICIO: Second, the inquirer fails to acknowledge the errancy of religious assertions. Whether disproved by science or rebuked from want of evidence, the unremitting refutation of religious claims belies the pretense of absolute and immutable truth. Thus, the second individual's contention of a youngish earth rests on quicksand. Such a declaration merely evinces the power of wishful thinking over cogent and independently verifiable evidence.

SAGREDO: Please, continue.

SIMPLICIO: Third, the questioner demands absolute proof for unwelcome assertions from science but offers none in return for his faith-based pronouncements. He possesses neither absolute proof nor a trace of evidence for his claims. He can only cite a private revelation from an unseen being or an artful interpretation from a Bronze Age text. Nothing more. His artifice serves merely as a blatant attempt to protect his faith, and his faith alone.

SAGREDO: Have you finished?

SIMPLICIO: Not yet. Fourth, and last, the interrogator's question itself is insightful. He likely was born into his faith, unquestioningly adopting the prevailing religion of his community. This accident of birth led to his indoctrination, from early childhood through adulthood, into that one faith. A charitable view recognizes he was reared to believe in absolutes, however ephemeral they might be. So, trained to think in this manner, the uncritical follower demands the same absolutes from science.

SAGREDO: Is there a less charitable view?

SIMPLICIO: Yes. A less benign view suggests a sinister motive. Perhaps the interrogator knows science cannot offer absolutes. Perhaps his demand for absolutes is merely a guise, an artful feint. As such, it constitutes a tainted form of *apriorism*. When confronted with loathsome scientific claims, his stipulation for absolutes permits him to both dismiss distasteful facts and retain his mystical beliefs. It is a skillful, clever dodge.

SAGREDO: Well said, Simplicio. Salviati could not have expressed it better.

SALVIATI: Indeed, I could not.

SIMPLICIO: Thank you both.

SAGREDO: Shall we meet again tomorrow for further discussion?

SALVIATI: By all means. Until tomorrow.

SIMPLICIO: Goodbye, Sagredo. Goodbye, Salviati.

SIMPLICIO: (to himself, while exiting) The journey to enlightenment is long, but with each step the fog of spiritual mysticism dissipates.

SAGREDO: (to himself) I am speechless.

SALVIATI: (to himself) I believe I have a tear in my eye.

---

Miscellany

Scriptural claims of a young earth derive, by definition, from sundry Biblical passages. Some proponents calculate the earth's age by counting the chain of genealogies from Adam to Jesus, estimating the lifespans of each individual in the chain. Others adjust their numbers from the slavery of the Israelites in Egypt, a period cited as 400[61] or 430[62] years. Still others decipher the planet's age from the reign of a king or from the construction of Solomon's temple.[63].

Five things not to say to a skeptic. Diagram 9.1

| What a Religionist Says | What a Skeptic Hears |
|---|---|
| (1) Scripture is the word of God; find truth in scripture. | *Circular reasoning*, where the conclusion is a restated form of the premise, erecting a self-verifying but fallacious argument. |
| (2) I prayed, and it came true. | *Post hoc fallacy*, where a conclusion is derived from the sequence of events, not their causality. |
| (3) The existence of God cannot be disproved. | *Negative fallacy*, where a claim is deemed true if it cannot be proved false; ignores that the onus lies with the claimant to prove a claim, not on the skeptic to disprove it. |
| (4) Beware false prophets. | *Apriorism*, where evidence that contradicts a strongly-held conviction is dismissed. |
| (5) God tells us so. | *Appeal to authority*, where the truth of a claim is asserted from the authority, prestige, or intelligence of the claimant, eschewing empirical evidence and testable predictions. |

# Ab Uno Disce Omnes

*The essence of religion is inertia; the essence of science is change. It is the function of the one to preserve, it is the function of the other to improve.*[64]
Winwood Reade (1838-1875)
British Historian

If a fracas arose between science as a way of thinking and science as a catalog of provisional truths, the prior would prevail. For it is, of course, the way of thinking that drives the catalog. It is the journey of curious minds, each pursuing the rigorous, transparent, and empirical path of science, that expands humanity's knowledge of the world. Said differently, it is the deliberate, resolute, and incessant trek from conjecture to confirmation (or refutation) that exposes new truths and, in equal number, overturns long-held fictions. Here, in chronological order, follows a selection of notable observations that capture the essence and operation of science.

*Knowledge should be his guide, not personal experience.*[65]
PLATO (C.428–C.348 BCE) GREEK PHILOSOPHER

*For disgrace lies not in imperfect knowledge but in foolish and obstinate continuance in a state of imperfect knowledge.*[66]
CICERO (106–43 BCE) ROMAN STATESMAN

*In order to examine into the truth, it is necessary once in one's life to doubt of all things.*[67]

RENÉ DESCARTES (1596–1650) FRENCH PHILOSOPHER

*The origin of all science is in the desire to know causes; and the origin of all false science and imposture is in the desire to accept false causes rather than none; or, which is the same thing, in the unwillingness to acknowledge our own ignorance.*[68]

WILLIAM HAZLITT (1778–1830) ENGLISH WRITER

*The mountebank takes advantage of this impatience in the human mind ... to tell his hearers: "This is what you are in search of; you now see and feel the truth, and need give yourself no further trouble about the mysteries of nature or the imperfection of your own faculties."*[69]

WILLIAM HAZLITT (1778–1830) ENGLISH WRITER

*The great tragedy of science is the slaying of a beautiful hypothesis by an ugly fact.*[70]

THOMAS HUXLEY (1825–1895) BRITISH BIOLOGIST

*Infidels are intellectual discoverers. They sail the unknown seas and find new isles and continents in the infinite realms of thought. An Infidel is one who has found a new fact, who has an idea of his own, and who in the mental sky has seen another star. He is an intellectual capitalist, and for that reason excites the envy and hatred of the theological pauper.*[71]

ROBERT GREEN INGERSOLL (1833–1899) AMERICAN LAWYER

*To dogmatism, the spirit of inquiry is the same as the spirit of evil.*[72]

AMBROSE BIERCE (1842–C.1914) AMERICAN SATIRIST

*Every great advance in science has issued from a new audacity of imagination.*[73]

JOHN DEWEY (1859–1952) AMERICAN PHILOSOPHER

*The terror of the thunderstorm led primitive man to the conception of a Supreme Being whose attribute was the thunderbolt. But when Franklin brought the lightning from the clouds and showed it to be a mere electric spark, when we learned to make the lightning harmless by the lightning-rod, and when finally we harnessed electricity to do our work, naturally our reverence for the thrower of the thunderbolt decayed. So the gods of experience vanished.*[74]

**CHARLES PROTEUS STEINMETZ (1865–1923) GERMAN-AMERICAN ELECTRICAL ENGINEER**

*Throughout the last 400 years, during which the growth of science had gradually shown men how to acquire knowledge of the ways of nature and mastery over natural forces, the clergy have fought a losing battle against science, in astronomy and geology, in anatomy and physiology, in biology and psychology and sociology. Ousted from one position, they have taken up another. After being worsted in astronomy, they did their best to prevent the rise of geology; they fought against Darwin in biology, and at the present time they fight against scientific theories of psychology and education. At each stage, they try to make the public forget their earlier obscurantism, in order that their present obscurantism may not be recognized for what it is.*[75]

**BERTRAND RUSSELL (1872–1970) BRITISH PHILOSOPHER**

*We need go back only a few centuries to find the great mass of people depending on religion for the satisfaction of practically all their wishes. From rain out of the sky to good health on earth, they sought their desires at the altars of their gods. Whether they wanted large families, good crops, freedom from pestilence, or peace of mind, they conceived themselves as dependent on the favor of heaven. Then science came with its alternative, competitive method of getting what we want. That is science's most important attribute. As an intellectual influence it is powerful enough, but as a practical way of achieving man's desires it is overwhelming.*[76]

**HARRY EMERSON FOSDICK (1878–1969) AMERICAN PASTOR**

*The truth is found when men are free to pursue it.*[77]

FRANKLIN D. ROOSEVELT (1882–1945) AMERICAN PRESIDENT

*A man ceases to be a beginner in any given science and becomes a master in that science when he has learned that ... he is going to be a beginner all his life.*[78]

ROBIN COLLINGWOOD (1889–1943) ENGLISH PHILOSOPHER

*My practice as a scientist is Atheistic. That is to say, when I set up an experiment I assume that no god, angel, or devil is going to interfere with its course; and this assumption has been justified by such success as I have achieved in my professional career. I should therefore be intellectually dishonest if I were not also Atheistic in the affairs of the world.*[79]

JOHN HALDANE (1892–1964) BRITISH SCIENTIST

*The four stages of acceptance:*
*1. This is worthless nonsense,*
*2. This is an interesting, but perverse, point of view,*
*3. This is true, but quite unimportant,*
*4. I always said so.*[80]

JOHN HALDANE (1892–1964) BRITISH SCIENTIST

*A common fallacy in much of the adverse criticism to which science is subjected today is that it claims certainty, infallibility and complete emotional objectivity. It would be more nearly true to say that it is based upon wonder, adventure and hope.*[81]

CYRIL NORMAN HINSHELWOOD (1897–1967) ENGLISH CHEMIST

*Science is a great game. It is inspiring and refreshing. The playing field is the universe itself.*[82]

ISIDOR ISAAC RABI (1898–1988) AMERICAN PHYSICIST

*Science is one of the very few human activities - perhaps the only one - in which errors are systematically criticized and fairly often, in time, corrected.*[83]

KARL POPPER (1902–1994) AUSTRIAN-BRITISH PHILOSOPHER

*Man masters nature not by force but by understanding. That is why science has succeeded where magic failed: because it has looked for no spell to cast on nature.*[84]

JACOB BRONOWSKI (1908–1974) BRITISH MATHEMATICIAN

*Representation of the world, like the world itself, is the work of men; they describe it from their own point of view, which they confuse with absolute truth.*[85]

SIMONE DE BEAUVOIR (1908–1986) FRENCH WRITER

*The scientific mind does not so much provide the right answers as ask the right questions.*[86]

CLAUDE LÉVI-STRAUSS (1908–2009) FRENCH ANTHROPOLOGIST

*It doesn't seem to me that this fantastically marvelous universe, this tremendous range of time and space and different kinds of animals, and all the different planets, and all these atoms with all their motions, and so on, all this complicated thing can merely be a stage so that God can watch human beings struggle for good and evil – which is the view that religion has. The stage is too big for the drama.*[87]

RICHARD FEYNMAN (1918–1988) AMERICAN PHYSICIST

*Science doesn't purvey absolute truth. Science is a mechanism. It's a way of trying to improve your knowledge of nature. It's a system for testing your thoughts against the universe and seeing if they match. And this works, not just for the ordinary aspects of science, but for all of life.*[88]

ISAAC ASIMOV (1920–1992) AMERICAN AUTHOR

*Physics is not a religion. If it were, we'd have a much easier time raising money.*[89]

LEON LEDERMAN (1922–) AMERICAN PHYSICIST

*If you want to save your child from polio, you can pray or you can inoculate.*[90]

CARL SAGAN (1934–1996) AMERICAN PHYSICIST

*[Belief in gods is a] worn out but once useful crutch in mankind's journey towards truth. We consider the time has come for that crutch to be abandoned. To say that 'God made the world' is simply a more or less sophisticated way of saying that we don't understand how the universe originated. A god, in so far as it is anything, is an admission of ignorance.*[91]

PETER ATKINS (1940–) ENGLISH CHEMIST

*Religion is a private pursuit of truth, driven by internal revelation and guided by authority. Science is the public pursuit of truth, guided by publicly executed experiments under controlled conditions. Religion is largely a culturally conditioned journey of sentimental wishful thinking. Science is transnational and transcultural. Religion is based on authority. Science is a ceaseless search for flaws in authority. You must choose between the feather bed of religion, which appeases the fears of the weak and encourages false hopes, and the rack of science, which exposes the true heart of the world.*[92]

PETER ATKINS (1940–) ENGLISH CHEMIST

*The invention of the scientific method and science is, I'm sure we'll all agree, the most powerful intellectual idea, the most powerful framework for thinking and investigating and understanding and challenging the world around us that there is, and it rests on the premise that any idea is there to be attacked. If it withstands the attack then it lives to fight another day and if it doesn't withstand the attack then down it goes. Religion doesn't seem to work like that.*[93]

DOUGLAS ADAMS (1952–2001) ENGLISH AUTHOR

*Every atom in your body was once inside a star that exploded. Moreover, the atoms in your left hand probably came from a different star than did those in your right hand. We are all, literally, star children, and our bodies made of stardust.*[94]

LAWRENCE KRAUSS (1954–) AMERICAN-CANADIAN PHYSICIST

*Science literacy is not only about what you know, it's about your depth of curiosity, and your capacity to evaluate evidence.*[95]

NEIL DEGRASSE TYSON (1958–) AMERICAN ASTROPHYSICIST

*Understanding that our understanding might be wrong is essential, and trying to figure out the ways we may be mistaken is the only way that science can help us find our way to the truth.*[96]

PHILIP PLAIT (1964–) AMERICAN ASTRONOMER

---

Miscellany

The Latin phrase *ab uno disce omnes* expresses the adage *from one, learn all*.[97]

# About The Author

*Critique the idea, not its proponent.*
Serge Bourgaine

The dialogues presented herein reflect a synthesis from an untold number of conversations in which the author, over the years, has engaged. In each instance, he voiced Salviati's perspectives to assorted Simplicio's and Sagredo's, listening, in turn, to their ideas and objections. In some cases, minds permitted the ingress of new paradigms; in others, less so. But all gained understanding if not agreement.

Readers of *Never By Authority*, an earlier text, will know Serge Bourgaine is a *nom de plume*. The rationale behind its use bears repeating. The author contends, perhaps too idealistically, that no apologia should derive validity from the standing, acumen, or fame (or lack of these qualities) of its proponent. The corollary, of course, is that claims and opinions should be evaluated solely from empirical evidence backed by testable predictions.

At risk of repetition, note this set of interrelated terms: empiricism, evidence-based thinking, scientific method, and inductive reasoning. All underlie a simple truth. They reflect a way of thinking, one free from dogmatic and mystical assertions. They accord an unrivaled ability to demarcate likely truths from wishful convictions. They comprise a method - one absent from the lion's share of human history and possessed by no other ideology - for reaching thoughtful, credible, and defendable conclusions. In short, they teach us how to think. From this method, Serge will believe anything; just provide cogent reasoning founded on empirical evidence and testable predictions.

*Serge*

# Epitaph

*In questions of science, the authority of a thousand*
*is not worth the humble reasoning of a single individual.*

Galileo Galilei (1564-1642)
Italian Physicist

# Endnotes

Entries that omit page numbers refer, in nearly all instances, to ebooks or similar digital content.

[1] Stewart 2008, not paginated.
[2] King James Bible Online, Psalms 1:2.
[3] King James Bible Online, Matthew 28:19-20.
[4] Mann 2011, not paginated.
[5] Pinker 2015, not paginated.
[6] King James Bible Online, Leviticus 18:22.
[7] King James Bible Online, Ephesians 5:3.
[8] King James Bible Online, Deuteronomy 7:1-26.
[9] King James Bible Online, Romans 3:28.
[10] King James Bible Online, James 2:24.
[11] King James Bible Online, Genesis 1:1-31.
[12] King James Bible Online, Jeremiah 13:1-5.
[13] King James Bible Online, Revelation 20:1-2.
[14] King James Bible Online, 2 Timothy 3:16.
[15] The Hadith, Sahih al-Bukhari 59:18.
[16] King James Bible Online, Genesis 3:1-4.
[17] King James Bible Online, Matthew 7:15.
[18] King James Bible Online, 2 Peter 1:20.
[19] Russell 1950, p. 74.
[20] Tilton 1940, pp 85-97.
[21] Gibbon 2003, p. 598.
[22] Laurie 2007, timestamp 00:29:25.
[23] King James Bible Online, 2 Peter 2:4.
[24] King James Bible Online, Mark 25:41.
[25] King James Bible Online, Matthew 18:8.
[26] King James Bible Online, 2 Thessalonians 1:9.
[27] King James Bible Online, Psalms 63:9.
[28] King James Bible Online, Proverbs 7:27.
[29] Bragg, The Earth's Core, 30 April 2015.
[30] King James Bible Online, Isaiah 14:15.
[31] King James Bible Online, Matthew 5:29-30.

[32] King James Bible Online, 1 Timothy 6:20.
[33] King James Bible Online, 2 Timothy 3:16.
[34] Tuchman 2014, not paginated.
[35] Sagan 2000, timestamp 00:01:25.
[36] Singh 1997, p. 32.
[37] Feynman 1994, p. 150.
[38] Russell 1950, p. 71.
[39] Bragg, Bird Migration, 6 July 2017.
[40] Allen 1980, timestamp 00:43:38.
[41] Keysar 2014, pp. 554-555.
[42] Roberts, Stephen. [www.askatheists.com/user/5972]
[43] King James Bible Online, Exodus 20:2.
[44] Harris 2005, timestamp 00:45:40.
[45] King James Bible Online, Genesis 1:1.
[46] King James Bible Online, Genesis 1:2-31.
[47] King James Bible Online, 1 Thessalonians 5:21.
[48] Dawkins 2008, paraphrased from interview.
[49] King James Bible Online, Genesis 10:25.
[50] King James Bible Online, Genesis 1:7.
[51] Boyer 2001, paraphrased from p. 320.
[52] King James Bible Online, Genesis 2:2.
[53] Lorenz 1965, not paginated.
[54] Ptolemy 2003, pp. 9James -12.
[55] King James Bible Online, 1 Chronicles 16:30.
[56] King James Bible Online, Psalms 93:1.
[57] King James Bible Online, Psalms 96:10.
[58] King James Bible Online, Psalms 6, 31, 37, 50, 101, 129.
[59] Norlind 1953, pp. 273-276.
[60] Austin 1983, p. 620.
[61] King James Bible Online, Genesis 15:30.
[62] King James Bible Online, Exodus 12:40.
[63] King James Bible Online, 1 Kings 6:1-38.
[64] Reade 1887, p. 34.
[65] Plato 2003, Stephanus 409, p. 337.
[66] Cicero 1960, pp. 174-175. Source text: *Non enim parum cognosse, sed in parum cognito stulte et diu perseverasse turpe est, propterea quod alterum communi hominum infirmitati, alterum singulari cuiusque vitio est adtributum.*
[67] Descartes 1911, principle 1, p. 219.
[68] Hazlitt 1934, p. 201.
[69] Hazlitt 1934, p. 201.
[70] Huxley 1898, paraphrased from p. 244.
[71] Ingersoll 1921, p. 309.
[72] Bierce 1911, p. 343.
[73] Dewey 1929, p. 294.
[74] Steinmetz 1922, p. 1089.
[75] Russell 1950, pp. 73-74.

[76] Fosdick 1941, pp. 140-141.
[77] Roosevelt 1938, p. 6.
[78] Collingwood 1947, maxim 1.46, paraphrased from p. 3.
[79] Haldane 1934, pp. vi-vii.
[80] Haldane 1963, p. 464.
[81] Bowen 1967, p. 536.
[82] Rabi 1964, not paginated.
[83] Popper 1965, p. 216.
[84] Bronowski 1956, p. 18.
[85] Beauvoir 1976, p. 242. Source text: *La représentation du monde comme le monde lui-même est l'opération des hommes; ils le décrivent du point de vue qui est le leur et qu'ils confondent avec la vérité absolue.*
[86] Lévi-Strauss 1964, p. 15. Source text: *Le savant n'est pas l'homme qui fournit les vraies réponses; c'est celui qui pose les vraies questions.*
[87] Gleich 1992, p. 372.
[88] Asimov 1988, p. 5.
[89] Lederman 1993, p. 198.
[90] Sagan 1996, not paginated.
[91] Atkins 1996, paraphrased from p. 6.
[92] Atkins 2000, paraphrased from pp. 25-31.
[93] Adams 2002, not paginated.
[94] Krauss 2012, p. 17.
[95] Tyson 2018, not paginated.
[96] Plait 2015, timestamp 00:01:05.
[97] Virgil 1893, p. 3, lines 65-66.

# Works Cited

The bibliography is organized by author surname and publication date.

Adams, Douglas. 2002. *The Salmon of Doubt: Hitchhiking the Galaxy One Last Time*. New York: Ballantine.

Allen, Woody. 1980. *Stardust Memories*. Voiced as character Sandy Bates. Metro Goldwyn Mayer. [video]

Asimov, Isaac. 1988. *Isaac Asimov - Part II*. Bill Moyers' World of Ideas. Show #130, 21 October 1988.

Atkins, Peter. 1996. Professor Says Science Rules Out Belief in God. *The Daily Telegraph*. Reported by Roger Highfield, 11 September 1996.

Atkins, Peter. 2000. Science and Religion: Rack or Featherbed, the Uncomfortable Supremacy of Science. *Science Progress*. Volume 83, number 1, Millennium Issue 2000.

Austin, David. 1983. The Trouble with Thinking Backwards. *New Scientist*. Cartoon accompanying article by Ralph Estling, volume 80, number 1360, 2 June 1983.

Beauvoir, Simone de. 1976. *Le Deuxième Sexe*. First published 1949. Paris: Éditions Gallimard.

Bierce, Ambrose. 1911. *Epigrams*. The Collected Works of Ambrose Bierce, Volume 3. New York: Neale.

Bowen, E. J. 1967. Sir Cyril Hinshelwood. *Chemistry in Britain*. Volume 3, pp. 534-536.

Boyer, Pascal. 2001. *Religion Explained: The Evolutionary Origins of Religious Thought*. New York: Basic.

Bragg, Melvyn. *In Our Time*. British Broadcasting Corporation (BBC). [audio podcast]

Bronowski, Jacob. 1956. *Science and Human Values*. New York: Harper & Brothers.

Cicero. 1960. *On Invention (De Inventione)*. First published 85 BCE. Loeb Classical Library, LCL 386. Translated by H. M. Hubbell. Cambridge: Harvard University Press.

Collingwood, Robin. 1947. *The New Leviathan or Man, Society, Civilization and Barbarism*. First published 1840. Oxford: Clarendon Press.

Dawkins, Richard. 2008. *Science and the God Delusion: A Conversation with Richard Dawkins*. UHI Millennium Institute. 2 April 2008. [video]

Descartes, René. 1911. *The Principles of Philosophy*. The Philosophical Works of Descartes, Volume 1. Translated by Elizabeth S. Haldane and G. R. T. Ross. Cambridge: Cambridge University Press.

Dewey, John. 1929. *The Quest for Certainty: A Study of the Relation of Knowledge and Action*. London: George Allen & Unwin Ltd.

Feynman, Richard. 1994. *The Character of Physical Law*. First published 1965. New York, The Modern Library.

Fosdick, Harry Emerson. 1941. *Living Under Tension: Sermons on Christianity Today*. New York: Harper & Brothers.

Gibbon, Edward. 2003. *The Decline and Fall of the Roman Empire, Volume 1*. First published 1776. Great Books of the Western World, Volume 37. Edited by Mortimer J. Adler, seventh printing. Chicago: Encyclopædia Britannica.

Gleich, James. 1992. *Genius: The Life and Science of Richard Feynman*. New York: Vintage.

*The Hadith*. First published circa 632 CE. [sunnah.com]

Haldane, J. B. S. 1934. *Fact and Faith*. London: Watts.

Haldane, J. B. S. 1963. The Truth About Death. *Journal of Genetics*. Volume 58, issue 3, December 1963.

Harris, Sam. 2005. *The View from the End of the World*. Seminars About Long-Term Thinking (SALT). 9 December 2005. [video]

Hazlitt, William. 1934. *Burke and the Edinburgh Phrenologists*. First published 1829. The Complete Works of William Hazlitt, Volume 20. Edited by P. P. Howe. London: J. M. Dent.

Huxley, Thomas Henry. 1898. *Biogenesis and Abiogenesis: The Presidential Address to the British Association for the Advancement of Science for 1870*. Discourses: Biological and Geological Essays, Volume 8. New York: D. Appleton.

Ingersoll, Robert Green. 1921. *The Great Infidels: A Lecture*. New York: C. P. Farrell.

Keysar, Ariela, and Juhem Navarro-Rivera. 2014. A World of Atheism: Global Demographics. *The Oxford Handbook of Atheism*. Edited by Stephen Bullivant and Michale Ruse. Oxford: Oxford University Press.

*King James Bible Online*. First published 1611. [kingjamesbibleonline.org]

Krauss, Lawrence. 2012. *A Universe From Nothing: Why There Is Something Rather Than Nothing*. New York: Free Press.

Laurie, Hugh. 2007. The Right Stuff. *House*. Voiced as character Gregory House, season 4, episode 2, 2 October 2007. Aired by National Broadcasting Company (NBC). [video]

Lederman, Leon, and Dick Teresi. 1993. *The God Particle: If the Universe Is the Answer, What Is the Question?* Boston: Houghton Mifflin.

Lévi-Strauss, Claude. 1964. *Mythologiques: Le Cru et Le Cuit*. Paris: Plon.

Lorenz, Konrad. 1965. When Man First Stood Up. *The New York Times*. Quoted by John Pfeiffer, 11 April 1965.

Mann, Charles C. 2011. *1493: Uncovering the New World Columbus Created*. New York: Vintage Books.

Norlind, Wilhelm. 1953. Copernicus and Luther: A Critical Study. *Isis*. Volume 44, number 3, September 1953.

Pinker, Steven. 2015. Why Free Speech is Fundamental. *The Boston Globe*. 27 January 2015.

Plait, Philip. 2015. Introduction to Astronomy. *Astronomy Crash Course*. Episode #1, 15 January 2015. Aired by PBS Digital Studios. [video]

Plato. 2003. *The Republic*. First published circa 380 BCE. Great Books of the

Western World, Volume 6. Edited by Mortimer J. Adler, translated by Benjamin Jowett, seventh printing. Chicago: Encyclopædia Britannica.

Popper, Karl. 1965. *Conjectures and Refutations: The Growth of Scientific Knowledge*. First published 1963. Second edition. New York: Basic.

Ptolemy. 2003. *The Amalgest*. First published circa 150 CE. Great Books of the Western World, Volume 15. Edited by Mortimer J. Adler, seventh printing. Chicago: Encyclopædia Britannica.

Rabi, Isidor Isaac. 1964. Humanistic Scientist. *The New York Times*. 28 October 1964.

Reade, Reade. 1887. *The Martyrdom of Man*. First published 1872. Twelfth edition. London: Trübner.

Roosevelt, Franklin. 1938. *Address at Temple University - Honorary Degree*. The Public Papers and Addresses of Franklin D. Roosevelt, Volume 5. 22 February 1936. New York: Random House.

Russell, Bertrand. 1950. *An Outline of Intellectual Rubbish*. Unpopular Essays. First published 1921. New York: Simon & Schuster.

Sagan, Carl. 1996. *The Demon-Haunted World: Science as a Candle in the Dark*. New York: Ballantine.

Sagan, Carl. 2000. Encyclopedia Galactica. *Cosmos: A Personal Voyage*. Episode 12. Aired by Public Broadcasting Service (PBS). [video]

Singh, Simon. 1997. *Fermat's Enigma: The Quest to Solve the World's Greatest Mathematical Problem*. New York: Walker.

Steinmetz, Charles. 1922. Religion and Modern Science: How They Differ, and the Place of Each. *Christian Register*. Volume 101, number 46, 16 November 1922.

Stewart, Ian. 2008. *Taming the Infinite: The Story of Mathematics*. London: Quercus.

Tilton, Eleanor M. 1940. Lightning-Rods and the Earthquake of 1755. *The New England Quarterly*. Volume 13, number 1, March 1940.

Tuchman, Barbara. 2014. *The Proud Tower: A Portrait of the World Before the War, 1890-1914*. First published 1966. New York: Random House.

Tyson, Neil deGrasse. 2018. *Twitter (neiltyson)*. 14 April 2018, 11:22 a.m. USA

CST.

Virgil. 1893. *Virgil's Aeneid, Book 2*. First published 19 BCE. Edited by J. C. Robertson. Toronto: W. J. Gage.

---

Miscellany

The quote that appears in the epitaph on page 97 is from François Arago, *The Biographies of Distinguished Scientific Men*, translated by W. H. Smith, Baden Powell, and Robert Grant. London: Longman, Brown, Green, Longmans & Roberts, 1857, p. 236.

*De omnibus dubitandum est*

ORBIS
UNUM
PRESS

An imprint of La Jolie Vie, LLC

Made in the USA
Monee, IL
08 November 2026

67972911R00069